# *That Terrible Texas Weather*

### *Tales of Storms, Drought, Destruction, and Perseverance*

Johnny D. Boggs

Republic of Texas Press

**Library of Congress Cataloging-in-Publication Data**
Boggs, Johnny D.
    That terrible Texas weather : tales of storms, drought, destruction,
    and perseverance / by Johnny D. Boggs.
        p.   cm.
    Includes bibliographical references and index.
    ISBN  1-55622-727-2 (pb)
    1.  Storms--Texas--History--19th century.   2.  Storms--Texas--History--
    20th century.   3.  Natural disasters--Texas--History--19th century.
    I.  Title.
    QC943.5U6 B64     2000
    551.55'09764—dc21                                99-049503
                                                           CIP

Republic of Texas Press is an imprint of Wordware Publishing, Inc.
No part of this book may be reproduced in any form or by
any means without permission in writing from
Wordware Publishing, Inc.

Printed in the United States of America

ISBN 1-55622-727-2
10 9 8 7 6 5 4 3 2 1
9912

All inquiries for volume purchases of this book should be addressed to Wordware
Publishing, Inc., at 2320 Los Rios Boulevard, Plano, Texas 75074. Telephone
inquiries may be made by calling:

(972) 423-0090

For the Sports Copy Desks
at the *Dallas Times Herald*
1984-1991
and the *Fort Worth Star-Telegram*
1992-1998

"If I owned Texas and all hell, I would rent out Texas and live in hell."

— General Philip H. Sheridan


"Everybody talks about the weather but nobody does anything about it."

— Mark Twain

# *Contents*

*Contents*

# Acknowledgments

It's impossible to thank everyone who helped me while researching this book, but here's a shot:

Margaret Farley of the Terrell County Historical Commission filled up a legal pad of coverage of the 1950s drought, plus provided information on the 1965 Sanderson flood. Susan Corbett and Mike Dougherty were valuable researchers/interviewers/folks with handy access to the *Fort Worth Star-Telegram* library. Others I must recognize include: Katie Anthony, Amarillo Public Library; Kim Lyerly, Austin History Center, Austin Public Library; Ellen Kuniyuki Brown, The Texas Collection archives, Baylor University; Patricia Herrera, Corpus Christi Public Library; Jeff Rogers and Denise Beeber, *Dallas Morning News*; Sharon Van Dorn, Dallas Public Library; Jeff Wilson, *Fort Worth Star-Telegram*; Mary Anne Welch, Goliad County Library; Barbara Wright, Gonzales Public Library; Kathy Stracener, Hearne Area Chamber of Commerce; Lisa Neely, Sarita Salinas, and Jamene Toelkes, King Ranch Archives; William H. Davis, National Archives and Records Administration; Janet Wall, National Climatic Data Center, Asheville, North Carolina; Shelly Henley Kelly, Rosenberg Library Archives, Galveston; Clark Bartee, U.S. Army Corps of Engineers, Galveston District; W.M. Von-Maszewski, George Memorial Library, Richmond; Clarissa Chavira, San Antonio Public Library; John F. Griffiths, Meteorology Department, Texas A&M University; John Anderson, Texas State Archives; Gillian Wiseman, Waco-McLennan County Library; and the staffs at Fort Concho National Historic Landmark, Panhandle-Plains Historical Museum, the Perryton Senior Citizens Center, and the Abilene, Angelo State, El Paso, Fort Stockton, Fort Worth, Kenedy, Midwestern State, San Angelo, and Santa Fe, New Mexico, libraries.

I would also like to thank Sarah Huffstetler of Arlington for the locator map that appears in this book; Elmer Kelton of San Angelo for his insight into the 1950s drought and his unequaled Texas

*Acknowledgments*

hospitality; Richard and Rita Sell of Perryton for their help with the Dust Bowl; my wife, Lisa Smith, for her editing and for putting up with my lengthy research trips across Texas; and my editor, Ginnie Siena Bivona, for her help and guidance.

# *Introduction*

Mike Harris, motorsports reporter for The Associated Press, sits in the press box at Texas Motor Speedway in Fort Worth. Clouds seem to threaten the running of the NASCAR Primestar 500 on this overcast March morning in 1999. The few sportswriters who have so far gathered in the press box debate on if, indeed, we will "get this race in."

Harris laughs as he recalls waiting in line downstairs, trying to guess what would be in store for the two hundred thousand spectators—weather-wise. A Texas sheriff had slowly turned around, looked Harris in the eye, and drawled:

"Son, there are only two kinds of people who predict the weather: fools and newcomers. Which one are you?"

Sure enough, the clouds dissipate, and nary a drop of rain will plague the stock-car race on this Sunday.

The tired old saying in Texas goes that if you don't like the weather, stick around a few minutes; it's bound to change. The saying really isn't a joke, either. A March 28, 1952, headline in the *Sanderson Times* tells the story: "Weather Continues in True West Texas Fashion—Uncertain." I got my first taste of the ever-changing Lone Star climate when I moved from South Carolina to Dallas to work at the *Dallas Times Herald* in 1984. I drove in to work one September morning sweating like a pig. When I left in the early afternoon to interview a high school football coach, I shivered in the freezing wind. A "norther" had blown in.

For more than fourteen years, I came to appreciate Texas weather. Like many other Americans, I learned the term "wind shear" after a thunderstorm caused a Delta Flight 191 to crash at Dallas-Fort Worth International Airport, killing 137 passengers and crew. Interviewing old cowboys, I realized that the "Dust Bowl" didn't just affect Oklahoma and Kansas.

In San Antonio, I learned that the slain Alamo defenders had been burned instead of buried because of a warm front that caused the corpses to ripen. And like other Texans, I froze, sweated, complained, and marveled.

I had scoffed at the idea of the omnipresent "baseball-size hail" until one spring evening in Fort Worth when I watched softball-size hail total cars in the *Fort Worth Star-Telegram* parking lot and leave dents in the sidewalk. I had believed that tornadoes don't strike major metropolitan areas until I came home to listen to my wife explain how she had spent the night in the bathtub with our wedding album while transformers blew, the wind roared, and our basset hound found a hiding spot under a living room table. (We won't mention the time I took the dog for a walk and got drenched before making it halfway around the block.) I saw people wash their cars in minus-zero wind chills, I drove on the freeway past submerged parking lots, and I watched straight-line winds snap trees like pretzel sticks. I tried to stay cool in the most miserable summers, slipped on ice during the most miserable winters, prayed for rain, and prayed for it to stop raining.

Eventually, I moved to New Mexico—but I swear it wasn't because of the weather.

This isn't a history of Texas weather. That would take years to research and volumes to write. Nor is it an explanation of why Texas has such meteorological extremes. I'll save that for climatologists and meteorologists. The best analysis of this phenomena is *Texas Weather* by George W. Bomar. The best year-by-year account is *One Hundred Years of Texas Weather 1880-1979* by John F. Griffiths and Greg Ainsworth.

Instead, this is a "sampler," if you will, of Texas weather through the years, a look at the storms and, above all, the people who endured.

"Why would anyone live there?" Western writer and Arizona resident John Duncklee asked me. He was referring to the tornadoes that devastated Kansas and Oklahoma in May 1999, but he could have been talking about Texas.

For an answer, I turn to Mrs. Ernestine Weiss Faudie, who recorded her story in Riesel, Texas, as part of the Federal Writers' Project, 1936-1940. After talking about the Indianola hurricanes, Faudie remarked:

"I have told you this true story of the coastal storms to show you what the old pioneers had to contend with, not only the pests of the insects on their crops, or the hardships of the lack of the comforts of life, but the very elements of nature, the drouths, the floods, and the unsettled condition of the country, even to desperadoes and murderers but never for an instant did we lose our faith in the future."

That faith still exists.

What follows are stories of Texas weather, but, above all, they are stories of a determined people with a will to overcome the elements.

## Key to Texas weather

 **Blizzard**

| | |
|---|---|
| Panhandle | 1886 |

 **Blue norther**

| | |
|---|---|
| Dallas | 1856 |

 **Droughts**

| | |
|---|---|
| South Texas | 1890s, 1950s, 1998 |
| West Texas | 1890s, 1950s, 1998 |
| North Texas | 1915, 1998 |

 **Floods**

| | |
|---|---|
| Ben Ficklin | 1882 |
| Hearne | 1899 |
| Sanderson | 1965 |

 **Hurricanes**

| | |
|---|---|
| Indianola | 1875, 1886 |
| Galveston | 1900 |
| Corpus Christi | 1919 |

 **Thunderstorm**

| | |
|---|---|
| Fort Worth | 1995 |

 **Tornadoes**

| | |
|---|---|
| Cedar Hill | 1856 |
| Goliad | 1902 |
| Austin | 1922 |
| Central Texas | 1930 |
| South Texas | 1930 |
| Waco | 1953 |
| San Angelo | 1953 |
| Wichita Falls, Vernon | 1979 |
| Saragosa | 1987 |

## Chapter 1

# A Wild May in Dallas County, 1856

### *"This is no earthly storm"*

Even native Texans wouldn't consider May 1856 your typical spring month in North Texas.

Across the United States that year, the debate about slavery began to heat up, resulting in violence. Pro-slavery forces attacked Lawrence, Kansas, in May, prompting John Brown and his abolitionists to kill five people favoring slavery in the Pottawatomie Creek region. Meanwhile, the African Methodist Episcopal Church opened a coed college in Wilberforce, Ohio. Indian wars continued against the Seminoles in Florida and the Yakima in the Pacific Northwest.

In Texas, people wondered about the weather.

Since 1854 the climate had become drier, but May and June continued to bring rain to Dallas County.[1] "Times are extremely dull money scarce and hard to obtain," Jonathan and David Merrill wrote from Dallas County in 1856. "Provisions scarce owing to two hard dry summers and the loss of our wheat crop a year ago by the Grasshoppers."[2]

The year 1856 had promised to be better, but May turned cold, ugly, and violent.

On May 4 a tornado struck the new village of Cedar Hill in the southwest Dallas County hills—more than thirty-three years before the first mention of a tornado in published national Weather Bureau summaries. The Merrills recalled the tornado in a letter as "one of the most Terafic [*sic*] Tornadoes in this neighborhood that is on Record." They said the storm hit April 29, but the letter was dated October 19, 1856, and other accounts say the funnel cloud struck in May.[3]

Jonathan and David Merrill cited casualties as nine killed and twelve severely wounded, with "every house torn to pieces, sills plates rafters and all broke into splinters or blown away. The largest sort of waggons tore to pieces the tires broken and twisted like bars of lead. Horses and Cattle carried from 1/4 to ½ mile waggons and teams flying throught the air. Such a mess of Distruction I never seen before. . . ."[4]

*Dallas Weekly Herald* editor J.W. Latimer compared the twister to the May 6, 1840, tornado that killed 317 people in Natchez, Mississippi, and a later storm that struck Vicksburg, Mississippi. "But we are of the opinion they were neither of them so terribly violent and furious as that which literally swept the village of Cedar Hill from the face of the earth."

Storm clouds swept in from the north and south, connecting about a mile south of Cedar Hill. The twister, more than one hundred yards wide, plowed up the ground, destroying fences and wagons and sending plows flying one-half mile. "Even the grass on the prairie was shorn off as with fire or scythe," the *Herald* reported. With intensifying fury, the storm moved on to the village. "Every house in the little village was razed to the ground and, except small portions of the foundations of some, blown away."

Henderson Hart sought shelter in the house of Mr. Berry, who was a partner in the Miller & Berry store, but the house was immediately "dashed to pieces—killing all the inmates except Henderson Hart, who is dangerously wounded." Mutilated dead bodies, including Mr. Berry's, "were scattered about in every direction." Berry's wife and child, along with a store clerk and another woman were also killed. The clerk, a man named Dickson, fled the house before the tornado destroyed it and grabbed hold of a post. Winds blew him around the post "until he was tied fast to it by his clothes. The post was then torn up and driven by all the fury of the storm to the ground and driven about with the body of the unfortunate young man fastened to it until every bone in his body was broken, and his flesh dreadfully mangled. When found, he was so securely bound to the post that he had to be disengaged by cutting his garments."

The family of John Hart became the storm's next victims. Hart, his wife, and their child died, with the child's body not found until the following day several hundred yards from the demolished residence. Also killed were Mr. and Mrs. Jacob Allen. Animals also suffered. Turning eastward, the storm destroyed more houses and churches, but no more fatalities were recorded. Four horses, twenty-three cows, nineteen sheep, and six hogs were killed around Cedar Hill, with a fence rail being driven through one ox. "Birds were killed in the air and found cleanly picked of their feathers." Meanwhile, timbers had been "torn into fragments, and driven into the ground with such force that it is said to be dangerous to walk over the ground at night for the sharp spars and splinters which cover the earth. . . ." Afterward, the living began the process of regrouping and rebuilding.

Survivors had been left without food or shelter—and many without clothes. Dallas County residents came to Cedar Hill to help remove dead stock, repair damages, and donate clothes and food. The dead were loaded onto a wagon and prepared for burial.[5]

Meanwhile, residents began recounting the strange stories that seem to happen in all tornadoes. The Merrills recalled how a boy had been carried nearly a half mile from his house, but survived. "One Lady lost the last stitch of her clothes when she was found and has since recovered and is doing well." A piece of silk from the Miller & Berry store was found the following day in Cedar Spring, twenty miles away. A hatbox, still containing its hat, had been discovered twenty-five miles away near White Rock Creek, and another hat was found on Rowlett's Creek, almost thirty miles away. "Dry goods and Boxes and hats were seen to fall about 50 miles from Cedar Hill in a N E direction."[6]

The *Herald* closed its account of the storm, saying:

> In view of the terrific fury of this dreadful tornado, surpassing in its power and terrible majesty anything we have heard of before, we are almost led to exclaim, with all reverence, however:
> > "The strife of fiends is on the battling clouds,
> > The glare of hell is in the sulphurous lightnings,
> > This is no earthly storm."

Things didn't get better in Dallas County, especially at La Reunion.

La Reunion, on the slope of a limestone hill three miles west of Dallas, had been settled in April 1855 by some two hundred French immigrants. Prime farmland this wasn't, and the colonists suffered from poor leadership and poorer weather.

In May 1856 crops fell victim when a blue norther—"a rapidly moving autumnal cold front that causes temperatures to drop quickly and often brings with it precipitation followed by a period of blue skies and cold weather"[7]—swept in from the northeast. As temperatures plummeted, the Trinity River froze solid and remained in that condition for three days.

Settlers shivered in their rough-hewn log cabins at the La Reunion settlement near Dallas. Some people called it quits.

"There was a scarcity of water for the cattle and livestock which caused great anxiety among the colonists," George Santerre writes in *White Cliffs of Dallas*. "Weeks and months of toil and planning had been wiped out overnight, the colonists were stunned and could hardly understand what had happened in this their promised land."

Others, however, remained. "Discouraged yet realizing that in order that they might live, they began the task of replanting the fields and their gardens."[8]

For the Reunion settlers, the weather remained uncooperative.

May passed, and June turned hot. As the corn and grain wilted, grasshoppers descended, destroying crops, grass, and tree leaves. The winter of 1856-57 turned savage. By the early 1860s the Reunion Colony had faded away. Today, La Reunion is remembered by a historical marker in the West Oak Cliff area of Dallas—not to mention downtown's Reunion Arena. Cedar Hill, however, rebuilt after the destructive tornado. A feeder trail to the Chisholm Trail passed through the village, as did the railroad. With the impoundment of Joe Pool Lake and population growth in Dallas County, Cedar Hill surged in the late 1980s and into this decade. It had a population of more than nine thousand in 1990.

Chapter 2

# The Indianola Hurricane of 1875

## "Send us help, for God's sake"

Cora Montgomery wrote in the 1852 book *Eagle Pass; or, Life on the Border*, "Everybody was disappointed in Indianola; it was so different from their ideas, but nobody found serious room for complaint." By 1875, however, few people complained about the Calhoun County town, except, maybe, rivals up the coast in Galveston.

Founded on Matagorda Bay in August 1846 as Indian Port, the town grew rapidly with its deep-water port and army depot. By 1852, three years after its name had been changed to Indianola, the town had become the county seat. Population soared to one thousand in 1860, and two thousand in 1870 as Indianola spread down the beach to Powderhorn Bayou. Railroad service started in 1871. Four years later the population had climbed to more than five thousand, and Indianola ranked second to Galveston as port cities in Texas.[1]

Business seemed better than usual on Wednesday, September 15, 1875. Visitors crowded the town because Bill Taylor stood on trial at the Calhoun County courthouse for the murder of William Sutton and Gabriel Slaughter. The violent Sutton-Taylor feud, which originated in DeWitt and Clinton Counties and involved notorious Texas gunman John Wesley Hardin, started as a disagreement over livestock in the 1860s. On March 11, 1874, William Sutton, along with his wife, Laura, and friend Slaughter, boarded the steamer *Clinton* at Indianola, bound for Galveston. Jim and Bill Taylor shot and killed Mr. Sutton and Slaughter, then fled. "It was assassination by cowards," Laura Sutton said. Bill Taylor had been arrested in Cuero, but Jim Taylor remained at large.

Winds out of the east increased during a day darkened by gray skies and the barometer dropped, but residents didn't seem worried. After all, gales were expected in mid-September. The tropical storm had first been recorded earlier that month east of the Lesser Antilles, then skimmed across Haiti and Cuba as it moved west through the Caribbean and into the Gulf of Mexico.

Texans weren't strangers to tropical cyclones in 1875. Five ships had been wrecked on Galveston Island on September 4, 1766. In 1818 pirate Jean Lafitte witnessed a storm also on Galveston that "raged with great violence," destroying four ships and turning two more into "dismantled hulks." An October 5, 1837 hurricane known as "Racer's Storm" (named after the British sloop) struck south of Brownsville, then moved over Corpus Christi and Galveston as it followed the Texas coast before heading across the Southeast and into the Atlantic. A strong hurricane on September 29, 1867, severely damaged Baghdad, Mexico, and Clarksville, Texas, destroyed wharves in Galveston, caused mudslides in Matamoros, and became known as the first "million dollar" hurricane in Texas. A lesser hurricane had even made landfall at Indianola on July 2, 1874. Indianolans were unafraid of the winds and rains on September 15, but things quickly changed.

As Thursday dawned, residents saw reason for concern. Waves broke over the beach, water covered the lower part of Main Street, and a heavy wind continued to howl. Boats transported people farther north in town. Businessmen first took currency and other valuables from their safes. Later they would simply try to save themselves. The barometer continued to drop while the water rose. By noon water "poured through the cross streets like a mill race." Buildings on Water Street began to succumb to the elements. Wharves were torn apart. Worse, roads leading out of town had become impassable. A passenger train sat beside the depot useless, its boiler drained the previous night and the cistern used to supply it contaminated with salt water. Those who remained in Indianola would have to ride out the storm.[2]

Businessman H. Seeligson tried to reach his home on horseback to rescue his family. He encountered water "saddle-skirt deep in many places, and a blinding rain of salt spray with heavy winds sweeping in perfect hurricane down the cross streets." Upon reaching his residence, he loaded his wife and children into a wagon, but after two or three blocks realized the hopelessness of his situation. "The waters were almost swimming in many places on main street; some places were full six feet deep and running at a fearful rate." As the winds increased, the Seeligsons took shelter in the courthouse.[3]

Others crowded into buildings they thought secure. As night fell, Seeligson watched as swells destroyed fences and small buildings. Larger buildings began to succumb. One building, housing thirty-one terrified people, was swept into Powderhorn Bayou. Only eleven escaped. Indianolans clung to debris. Others tried to build improvised rafts. One man saved himself when he was washed into the wreckage of a depot building, only to watch his sister—almost within his reach—disappear under the water. Parents moving through the riverlike streets hoisted their children overhead, but waves swept many of the youths to their deaths.

In the courthouse, water burst through the east door, forcing the Seeligsons and other refugees to move to the second story, where they spent a frightful night. Children slept in makeshift beds. Wrote Seeligson: "The building, although constructed in a [splendid] manner of the stoutest masonry, with foundation six feet deep and five in diameter, rocked as though an earthquake was in progress. The rushing of the waters through the lower doors and windows and the wild [fury] of the storm . . . was so deafening that one could scarcely hear his own voice."[4]

C.A. Ogsbury, editor and publisher of the Indianola *Bulletin*, along with his wife, two children, and mother-in-law, took cover in R.D. Martin's two-story house. They also were forced upstairs by rising water. "Night coming on, the situation was awful. Screams from women and children could be heard in every direction," Ogsbury said.[5]

William Coffin had stayed in his house with his wife and two children. The house collapsed, and water rushed in. The children drowned, Mrs. Coffin died—"of exhaustion," according to the *Advocate*; killed instantly when the house split apart, according to historian Brownson Malsch—and Coffin grabbed a piece of timber and finally drifted to land, "where he watched by his dead wife until the storm was over." One of his children's bodies was found six miles west of Indianola.[6]

"Numerous instances of heroic struggles with the current are told," the *Victoria Advocate* reported. Mrs. Ernestine Weiss Faudie recalled one that took place in the courthouse, involving Bill Taylor and another prisoner: "They had been placed in the court house and during the height of the storm both frequently swam [through] the court house windows to rescue some drowning person." It's a good story, but it's undoubtedly just a tall Texas tale. Faudie wasn't in Indianola at the time and told the story more than sixty years after the hurricane. Indianola historian Brownson Malsch records a more plausible account: Prisoners Bill Taylor, George Blackburn, and Sam Ruschau gave their word that they wouldn't try to escape and were released from jail when the water level threatened their safety. Instead of keeping their pledge, they stole horses, made it across the flooded prairie, and left word at Green Lake where the owners could find their horses, "thus evincing some sense of honor and decency."[7]

Some people stabilized their homes by chopping holes in the floors, allowing the water to enter. John Garner and his family, and others, took shelter in a two-story home, then saved themselves and the house by leading twenty horses into the first-floor rooms. The weight of the animals kept the house secure as the lower floor flooded. The refugees survived, but the horses drowned.

Between 9 P.M. and midnight, the wind changed direction, pounding the town from another front. "Buildings were hurled against each other, adding wreck to ruin until the waves were fairly covered with timber from the ruined homes," Seeligson said. "Roof after roof was swept by...." One four-year-old girl somehow climbed onto a floating roof, wedged her fingers underneath the shingles, and held fast. She was rescued fifteen hours later after the roof structure had been deposited on dry land.[8]

One humorous account of the storm on Mustang Island comes from the Mercer family diary: "Joe Hall lowered the whiskey bottle 3 inches at one drink and Barnes took 7 long swallows."[9]

There was nothing funny in Indianola, however.

"Many were forced out of the second stories when the water rose in them and had to seek safety in hastily constructed rafts which they made from the sections of the floors and walls of the houses they were in," Faudie recalled. "Some of them were thoughtful enough to have ropes and they were lashed to the rafts by them, but many were drowned when the buildings they were in collapsed and the people were crushed or drowned."

By dawn Friday the storm had subsided. At 6 A.M. the flood-waters had receded, but the horror was far from over. Indianola, the booming port town, had been turned into "a ruined City!"[10]

Bodies were found for twenty miles along the shore, and only eight buildings had escaped damage from the storm. "It is difficult even to identify where many of the buildings stood. . . ." the *Advocate* reported. Nor had the hurricane limited its destruction to Indianola. Ninety percent of the residents of Upper Saluria drowned. Galveston went through a "terrible state of affairs," trees and crops were leveled in East Texas and Louisiana, tornadoes formed, and Austin County was deluged with wind and rain. In Indianola, parts along the beaches of Matagorda Bay and Powderhorn Lake were littered with the wreckage of buildings, "wooden cisterns, pianos, trunks, boxes, barrels, chairs, tables, sofas" and the remains of dead animals.[11]

Search parties looked for survivors. Entire families had been wiped out.

"My brother-in-law, his family, and his home had disappeared and were never seen again," Ernestine Weiss Faudie recalled, although "my husband hoped for months to hear of him." Faudie's in-law, Henry Hamburg, "was in charge of the Methodist church at this place when the big storm came. . . ." The Advocate listed a Reverend Hornberg, with wife and child, among the missing and the German Methodist Church as "gone."

The *Advocate* became the first newspaper to detail the destruction. A reporter arrived on horseback at 9 A.M. Sunday after some twelve hours in the saddle. An extra edition was issued on September 24, with the headline:

Indianola!
THURSDAY'S STORM!
A DAY OF DANGER
AND NIGHT OF
HORROR!

Settlers in Cuero, Victoria, San Antonio, and elsewhere began sending relief parties. Aid later came from Galveston, New Orleans, and Corpus Christi and as far away as Detroit, New York, and Charleston, South Carolina. Seeligson recalled that aid from "foreign sources" totaled $15,000, although "much suffering remains to be relieved." At Galveston, news of the devastation came aboard the steamer *Harlan*. Indianola District Attorney W.H. Crain sent a plea for help via the ship to a Galveston newspaper: "We are destitute. The town is gone. One-tenth of the population are gone. Dead bodies are strewn for twenty miles along the bay. Nine-tenths of the houses are destroyed. Send us help, for God's sake."[12]

Burial parties began forming Saturday morning, and food and water arrived from Victoria on Monday. Buildings were reported as "gone, destroyed, badly scattered, wrecked, badly shattered, moved southward, swept away...." The *Advocate* noted that "Across the bayou but one building remains standing, a two-story frame used as a hotel." The courthouse "escaped without serious damage," but only the Presbyterian church was left standing. The Catholic church "was crushed together," the other four or five churches were gone, and the Masonic building had been "washed away."[13]

Estimates of the dead ranged from a low of one hundred seventy-six to a high of eight hundred, although the latter figure would include figures from the storm's entire path. The *Advocate*'s September 24 extra estimated one hundred bodies recovered and one hundred fifty-eight missing. Malsch writes that the death list grew to two hundred seventy, with several bodies never recovered. Ships discovered the dead bodies of animals and humans in the Gulf of Mexico and Matagorda Bay for days after the storm. There were also reports of "robbers pillaging the dead" and fifteen of the looters killed on sight. Plus, the *Paisana* was lost somewhere between Brazos Santiago and Galveston; the side-wheel steamer had reportedly been carrying $200,000 in canvas bags.

Other ships also suffered. The *Advocate* reported that several sloops and schooners had been beached, wrecked, or lost. At least one schooner had been hurled five miles inland.

Although newspapers reported articles and the hurricane's path has been charted, Texas meteorologist George W. Bomar writes that few scientific details of the storm are known. Indianola had had an official weather observation and reporting station since 1872, and Sergeant C.A. Smith manned his post at the station during the storm as long as possible. At 5 P.M. September 16 Smith recorded the barometer at 28.90 corrected, the temperature at 75 degrees, and an anemometer reading of eighty-two miles per hour. The anemometer blew away fifteen minutes later in winds at eighty-eight miles per hour. Gusts were later estimated as high as one hundred fifty miles per hour. At 6 P.M. a warehouse behind the observation station collapsed, forcing the evacuation of the station.

Men pose with wreckage left by a hurricane in Indianola, in 1875.
*Texas State Library & Archives Commission*

Slowly, some survivors in Indianola began to regroup. Others moved away. Those who remained rebuilt, but the town never came close to reaching its peak of 1875. Seeligson wrote of his hopes that the tragedy would serve as a warning, but no one seemed to listen. City officials tried to move the new Indianola to higher ground, perhaps three and a half miles above the bayou's mouth, but the money-conscious Morgan Steamship Line refused to cooperate, so the town remained in harm's way. The Morgan decision prompted several leading merchants, including the *Bulletin*'s C.A. Ogsbury, to move. Those who remained gambled that the hurricane of 1875 had been a "once-in-a-lifetime" storm. Their gamble would prove costly when a second such storm destroyed Indianola in 1886.

Chapter 3

# The Ben Ficklin Flood of 1882

## *"So terrible a calamity"*

It's hard for many people who don't live out West—and for some who do—to picture just how devastating a flash flood can be. Heavy rainfall in a short amount of time can send a wall of water destroying anything in its path. George W. Bomar points out in *Texas Weather* that Texas is the site of destructive floods every year. Those floods often claim many lives and result in massive property damage, reaching thousands of dollars and sometimes millions of dollars.

Such was the case at Ben Ficklin (sometimes spelled Benficklin) on August 23, 1882.

In west-central Texas, Major Benjamin F. Ficklin bought six hundred and forty acres on the South Concho River and set up a San Antonio-El Paso stagecoach station in the late 1860s. In 1873 Francis Corbett Taylor, William Stephen Kelly, and Charles B. Metcalfe laid out a town nearby and named it after Ficklin, who had died two years earlier. A post office was established on August 27, 1873, Tom Green County was organized the following year, and in January 1875 Ben Ficklin became the county seat, defeating Santa Angela (which became San Angela and later San Angelo) at the polls.[1]

Ben Ficklin and Tom Green County boomed. By 1880 the county had a population of more than thirty-six hundred and post offices at Ben Ficklin, Fort Concho, Knickerbocker, and San Angela. Ben Ficklin, which boasted a population of six hundred in 1879, had a three-room jail by 1878, a two-story stone courthouse by February 1882, and a rivalry with nearby San Angela.[2]

San Angela was bigger than Ben Ficklin but had a ne'er-do-well reputation as "a horridly wicked portion of the universe,"[3] a parasite town to suck money and blood from the soldiers at nearby Fort Concho. San Angela tried for the county seat in 1878 but lost again. Things continued to look bright for Ben Ficklin.

On Wednesday evening, August 23, 1882, it began to rain. "It came in torrents for more than an hour," the *Tom Green Times* reported, stopped, then started again. Thunder roared and lightning streaked across the sky. "The ceaseless roar of water from roofs and gutters made old Texans anxious for to-morrow and what to-morrow would reveal."[4]

The year 1882 had been a wet one in Tom Green County. Rains had fallen hard since May. Charles B. Metcalfe recalled in 1924 that "So much water had fallen the draws were flowing clear water everywhere. The country slopes, causing a current of irresistible power, which in flood time uprooted trees, changed channels, and obliterated houses or any movable objects it met. As the country had been soaked for weeks it was of course already full of water" by August 23.[5]

Wrote historian Susan Miles: "Barrels, buckets, vats and tubs, often so heartbreakingly empty, now became reflecting pools for the passer-by."[6]

Rainfall that night measured 6.9 inches, and Dove Creek, Spring Creek, and the Middle and South Concho Rivers flooded. Water began rising in Ben Ficklin the morning of the twenty-fourth. By 11 A.M., the water height reached forty feet.[7] "Such a rise has not been known to the oldest dweller on the frontier," the *Times* reported August 26 "and Heaven grant such another will not be seen by this generation."

Floodwaters, the paper went on, were filled with "chairs, goods, trunks, boxes, and furniture of all kinds ... hundreds of sheep ... roofs of houses, quantities of planks...." Trees were uprooted "and hurried along as if mere toys in the fierce element."

The flood took a heavy toll in human life, also. The *Times* devoted almost the entire front page of its August 26 edition with coverage of the disaster. Its headline blared:

TERRIFIC FLOOD!
GREAT DESTRUCTION
OF LIFE.
Thousands of Dollars Lost in the Seething Waters.
BEN FICKLIN ALMOST WHOLLY RUINED,
AND SAN ANGELA PARTIALLY
INUNDATED.
Some Particulars of the Disaster.

As the waters rose, people began to flee to higher ground by any means possible.

F.A. Karger made his way to safety "and watched the water carry away all of his worldly goods." Later he heard gunshots and realized county surveyor H.B. Tarver was trapped in the courthouse and was shooting his gun to grab someone's attention. Karger and others tried to rescue Tarver in boats but couldn't get to the courthouse because of the swift current. Tarver had to wait out the flood for twenty-four hours in the building, which was not destroyed.

Thirteen prisoners, in jail on charges from drunkenness to murder, were rescued by Clint Johnson and others "when the water was pressing against the door." Johnson climbed into a tree. Bit Martin, who had built a hotel in Ben Ficklin, was warned by a relative of the danger. Martin escaped with his wife and four daughters on a hack despite waters reaching the bed of the wagon.

Mrs. Cicero Russell recalled that her father, John Burleson, rescued a boy during the flood. The Burlesons had moved to the country in 1875, ran a dairy farm awhile, and lived just below Ben Ficklin. The boy, Cliff Gill, "was about eight or nine years old, I think," Russell said. "He lived with us until he married."[8]

At the old stagecoach station, C.D. Foote had driven from town and convinced several people living there to leave. Mrs. M.J. Metcalfe, however, declined, saying she thought the water would not continue to rise. Metcalfe's daughter, Zemula, also decided to stay. "My place is with mother," she said. When the water continued to rise, the Metcalfes had a change of heart and boarded S.C. Robertson's hack and tried to make higher ground. The horses "balked," however, and so the settlers returned to the station and climbed onto the roof of the Metcalfe home.

Sometime later, Terrell Harris and Kerby Smith arrived in "a frail boat." The Metcalfes, Robertson, Blake Taylor Sr., George Robinson, a Mexican, and a black cook named "Red" Evans waited anxiously, but the boat, half full of water, capsized ten yards from the house, and Harris and Smith swam to a nearby grove of pecan trees and climbed up. Harris almost didn't make it. He was caught underneath the boat after it capsized. He raised the boat off, climbed on top, and pulled off his boots before swimming to the pecans. From their perch in the pecans, Harris and Smith watched in horror as the Metcalfe house's roof broke into, the men on one side and women on the other. Taylor lost his hold and was swept away. Meanwhile, "The ladies bore down upon the pecan clump, and were [engulfed], screaming as they disappeared...."

Robertson fell from the mangled roof, but took hold of a tree and "held on through as terrible a day and night as any one ever endured." Wreckage from the house knocked him off his perch, but he swam to safety and was rescued from another tree shortly after 6 A.M. Friday. Harris and Smith, "bruised and sore from contact with the drift and from their super-human exertions," were rescued about an hour later. Of those who took refuge on the Metcalfe roof, only Robertson survived. Mrs. Metcalfe's body was found sixteen miles away at the mouth of Crow's Nest. She was buried on August 26, the day George Robinson's body was discovered.

"The station is itself an utter ruin," the *Times* reported. "All the houses and walls have disappeared in a current before which stone walls melted like snow in June."[9]

The destruction wasn't limited to Ben Ficklin. At San Angela, water "rose more than two feet above the first gallery" of the $11,000 Concho House, which was rendered "wholly unfit for further use." The *Times* reported that "All adobe houses touched by the water are ruined." Homes and businesses were washed away, and the *Times* originally overestimated the damages in the city at $50,000 before reducing the figure to $15,000. In the surrounding county, "Hundreds of carcasses of sheep and other stocks are scattered over the prairies, wagons, blankets, clothing and all sorts of furniture strew the banks...." Telegraph lines were down.[10]

Ben Ficklin, however, suffered the most.

"Such sickening details and horrid suspicions of death we do not often meet in life," the *Times* reported on August 26. "Our reporter is out hunting up the news, and in our next issue we will give full particulars of the disaster. As it is we feel a relief in closing the initial chapter of so terrible a calamity."

Although the *Times* originally estimated that "not fewer than one hundred lives were lost in this county," the death toll reached only about sixty-five. (The September 2 *Times* listed "a total of 53 persons lost, so far as known.") "A Mexican lost his wife and five children just opposite" Ben Ficklin, the newspaper reported. "Twenty persons are missing from Mr. W.S. Kelley's ranch, on Spring Creek, and ten from [Richard F.] Tankersley's. Three at Mr. David McCarthy's . . . are lost, and all his sheep are gone. Dr. Owens and child, living on Levy's ranch, are drowned, and his wife was taken from a tree on Friday morning at the same place. Mrs. H.K. Mathis was swept away from her home with a child in her arms."

Telegraph lines carried the sad news to families. Joseph Spence telegraphed his father in Austin: "Mrs. Metcalfe and Zemmie drowned—everything lost—rest of family and Robert safe." J.B. Taylor Jr.'s wire read: "Pa is no more; Josie is safe."[11]

Search parties began forming that Friday.

"I saw many of the victims brought out of the water," Becky Sanford recalled in 1938. "Some were hanging in the trees and others were washed away downstream."[12]

The body of a girl about six years old was discovered by a Colorado man that Sunday near the mouth of Crow's Nest and was "decently interred." Station cook "Red" Evans's body was found near the post lime kiln.

In Ben Ficklin, residents began assessing the damage. All that remained on the flat were the courthouse, jail, and two houses, and the *Times* noted that the courthouse "is injured to such an extent that it will probably fail." On the hill, fifteen houses and the schoolhouse still stood. The thirteen prisoners spent the following three weeks outdoors under guard—and helping with the search—before they could be moved back into the jail.

The September 2 *Times* estimated damages at:

| | |
|---|---|
| Dove Creek | About  $10,000 |
| San Angela | $15,000 |
| Spring Creek | About $30,000 |
| Miscellaneous losses | $15,000 |

Losses at Ben Ficklin and vicinity, however, totaled about $115,000, according to the newspaper.

A reporter "found the upper part of the town more to represent a camping ground than anything else. Tents were pitched in different places and bedding of every description was strewn on the grass and fences."[13]

But the settlers, even Ben Ficklin's rivals in San Angela, quickly responded. "Tom Green county may boast of having the most sympathetic citizens in the world," the *Times* lauded. "Directly the sad news was known, parties from all sections of the county came forward and offered every assistance ... Messrs. Foote, Lackey, and W.H. Lessing opened their houses and said 'take what we have.'" In San Angela, a meeting was held on August 28 to aid the victims of the Ben Ficklin flood. More than $4,000 came from San Angelans, while gifts also came from residents in Mason, Colorado City, Loyal Valley, Fort Worth, Fredericksburg, Fort Davis, and businesses in San Antonio, Galveston, and Fort Worth. The Sixteenth Infantry Band of Fort Concho gave a benefit concert in San Angela. The concert was "a grand success," the *Times* reported September 2, "and the music discoursed was excellent." In the end, $5,256.50 was reportedly raised for the Ben Ficklin survivors.

Although some settlers remained, the town of Ben Ficklin was history. Some survivors moved to Sherwood, southwest of San Angela, while others opted for San Angela, where jobs and free homesites could be found. San Angela also took in the county offices and post office.

The September 16 *Times* noted: "Our streets have been lined with freight wagons, and vehicles of every description bringing people into San Angela to make their purchases. Lumber has commenced to arrive and many buildings are being erected."

Notices appeared in the September 2 *Times*, reporting change of venues: "John Engel, late of Ben Ficklin, would respectfully inform his friends and the public that he has removed from the washout, and may now be found at Keyser's old store on Oakes Street." "W.H. Brown Late of Ben Ficklin would respectfully inform his friends, also the public in general, that he has removed to San Angela and can be found at the Blue Ribbon Saloon. . . ." The Tom Green County Land Agency's C.D. Foote noted that because "an unprecedented calamity overwhelmed the village of Ben Ficklin . . . as soon as my office can be fitted up, I will resume business in the town of San Angela, where my clients will please address me."[14]

The first seat of Tom Green County had been washed away. San Angela had finally bested its county rival, with help from Mother Nature. Voters moved the county seat to San Angela, which became San Angelo, in 1883. The following year, the San Angelo courthouse was completed. The town continued to grow as an agriculture, education, military, and tourism center. The Chamber of Commerce estimated the city's 1998 population at 93,000, dominating Tom Green County's estimated population of 103,000. And Ben Ficklin? A Texas Historical Commission marker, erected in 1965 at the town site, is all that remains.

"I guess it was so thoroughly devastated that people thought if it could happen this time, it could happen again," says novelist Elmer Kelton of San Angelo. "So most of the people moved down the road to San Angelo.

"Over the long haul, however, the Ben Ficklin flood was the impetus for the building of dams in the area."[15]

## Chapter 4
# The Blizzard of 1886

**"That was the worst winter I ever saw"**

To understand the terrible blizzard that swept across the Texas Panhandle on January 7, 1886, first you need to go back a few years.

By the early 1880s, with the surrender of the Comanches and Kiowas and the death of the Apache warrior Victorio, the Indian wars had ended in Texas. The buffalo had nearly been exterminated, and the cattle industry boomed. Across the West, men and women lauded the wealth of beef. It was like another gold rush. Go West, raise cattle, make a fortune. Books were published glamorizing the industry. James S. Brisbin's *The Beef Bonanza; or, How to Get Rich on the Plains* came out in 1881 followed by Major W. Shepherd's *Prairie Experiences in Handling Cattle and Sheep* in 1884 and Walter Baron von Richthofen's *Cattle-Raising on the Plains of North America* in 1885.

von Richthofen called cattle-raising "a legitimate and safe business," estimating average losses as two to three percent annually from sickness, railroad accidents, straying, theft, and "extreme cold with lasting snow." About the hard winters, especially in Wyoming and Montana, he wrote:

> The prairie is covered then by a crust of ice, and the cattle are not able to get enough food. In such extreme cases the poor, lean cattle, which are too weak to endure several days' fasting, die of hunger. This happens very seldom, as such storms are not frequent in this climate.[1]

Things looked like Utopia when Baron von Richthofen wrote his book. "Never in the history of the cattle business of the West has the future of this industry looked brighter and more promising than at the present time."[2]

By 1888, however, many ranchers had different feelings. The drought of the 1880s and the hard winters that followed bankrupted many and left thousands of cattle dead on the plains.

Western cattle ranchers got a look at winter perils during the winter of 1880-81. Blizzards on the high plains sent cattle drifting south. The animals would turn their backs to the wind and move like ghosts. Nothing would stop them. By spring, thousands of cattle from as far north as the Arkansas and Platte River rangers were found grazing in the Texas Panhandle. Cowboy crews came from Kansas, Nebraska, Colorado, and Wyoming to find their cattle on the Canadian and Red Rivers, where the grass had been picked clean. Texas cattle had drifted south, too. Panhandle ranches sent cowboys to the Blanco, Brazos, and Colorado Rivers. Many worked as "floaters," consisting of a chuck wagon and a few men who would go around and look for strays and bring the cattle back north to their home ranges.[3]

The incident left Texas cattlemen concerned that the northern cattle would ruin their winter pastures when the Panhandle was already overstocked. They turned to barbed wire.

Joseph F. Glidden had patented his barbed wire on November 24, 1874, and the wire was being strung up in Texas by 1875. The wire had been designed to keep animals out, and many Panhandle pastures had been fenced in by the early 1880s to protect grass, water rights, and cattle herds.

The T Anchor Ranch enclosed two hundred and forty thousand acres in 1881-82. Posts were put eighty feet apart so that when an antelope or mustang hit the fence, the strands would bend but not break. Charles Goodnight fenced in his Quitaque Ranch in 1883 and the Tule Ranch in 1884-85. Profit-minded freighters would bring in barbed wire and leave with buffalo bones to sell. Most fences were put up with four barbed-wire strands and cedar posts two rods apart, although some fences used three or five strands, even more, and some posts were closer together. Posts usually came from the Canadian River breaks, No Man's Land, and Palo

Duro Canyon. Cost of the fences ranged from $200 a mile to $400 a mile.

Ranchers quickly learned that the barbed wire would keep northern cattle off their ranges. In 1882 the Panhandle Stock Raisers Association met in Mobeetie, Texas, and decided to begin building a drift fence from the New Mexico Territory line to the Canadian River breaks in Hutchinson County.

Judge O.H. Nelson told noted Texas historian J. Evetts Haley in 1927:

> We built the drift fence to keep the cattle from drifting south. They did not drift north; they only drifted during the storms in the wintertime and they would go as far as the Pecos without stopping.[4]

Northern Panhandle ranches such as the LE, LIT, LS, LX, Turkey Track, Bar CC, 7K, and Box T joined forces in building the drift fence. Individuals also put up wire. Charles Goodnight built a sixty-mile-long drift fence from near the Armstrong County line, between Salt Fork and Mulberry, to Coleman and Dyer's Shoe Bar range. Goodnight's fence connected with the Shoe Bar wire that stretched to near Memphis. Together, the Goodnight-Shoe Bar drift fence stretched one hundred miles.

The fences, put up by different ranchers with different ideas, didn't always connect, but few cattle could find their way through this maze. The Panhandle ranchers were pleased by how well the fences protected the Canadian River breaks. By 1885 drift fences crossed the Panhandle and thirty-five miles past the New Mexico Territory line.

But what about gates?

Laws enacted in 1884 and 1887 said fences must have gateways. The 1887 law required gateways be at least ten feet wide on any fence line "more than three miles lineal measure running in the same direction."[5]

During the severe winter of 1884-85, the drift fences kept the northern cattle out of the Panhandle grasslands, but with a high cost.

Cattle drifted to the fence and wouldn't move. They huddled together along the barbed wire and froze to death. The Panhandle drift fences became "a chain of carcasses." Along the Goodnight-Shoe Bar fence, antelope huddled in a fence pocket where Clarendon settlers killed some fifteen hundred of them. The LX Ranch lost three hundred cattle, and the blizzard pushed twenty-five thousand to thirty thousand cattle drifting from the Northern Plains. The JA Ranch sent cowboys to water drifts and push the cattle back onto the plains. By early spring, Panhandle cattle were fat on good grass.

The drift fences had saved the range. Yet the ranchers didn't seem to understand just how dangerous the lines of barbed wire could be. As the *Fort Worth Daily Gazette* reported on February 27, 1885, ". . . fences, short grass, poor and dying cattle, and hard times made their appearance about the same time."[6]

Texas ranchers would learn a bitter lesson in 1886.

Wednesday, January 6 had been mild in parts of the Southern Plains, but a cold front swept through that night, bringing heavy snow and high winds. By Thursday the brunt of the storm was pounding Texas. Buffalo Springs was covered with two to three feet of snow; the road to Trinidad, New Mexico Territory, was blocked for a month. Near Mobeetie, winds were measured at fifty-eight miles per hour at Fort Elliott. The Panhandle wasn't the only region freezing. Temperatures dropped to minus seven degrees in Dublin, minus six and a half in Austin, zero in Palestine, and eleven in Galveston. By 9 P.M. ice had formed on Galveston Bay. The bay would be frozen three to four inches in a few days, and six inches of snow would fall on the port city. Indianola and Brownsville also felt the chill.

"It was plenty cold," LX Ranch cowboy John Arnot remembered. "Cowboys found frozen antelope that winter."[7]

On January 8 the temperature at Fort Elliott was minus ten degrees. On January 13 the *Chicago Daily Drovers Journal* reported bad news from Texas: "Intelligence from the cattle regions of the state are of the gloomiest and most depressing character. The loss in cattle will be great."[8]

Cattle in western Kansas, present-day Oklahoma, and the Texas Panhandle totaled two and a half million during the winter of 1885-86. The animals panicked in the howling wind and blowing snow, turned south, and started their "blizzard gait," a virtual stampede until they were stopped by the long lines of barbed wire fences.

Cowboys went to work, trying to save the cattle. Recalled one:

> Eleven of us camped on the [drift fence] in the winter of 1885 and 1886. That was the worst winter I ever saw. The boys were from different outfits: two LS men, two ZH, two OI, two OX, and others. All but the LS men were sent down by the northern outfits to look after their cattle that drifted. We didn't do much good it was so damn cold.[9]

Another cowboy had been horse hunting in the Panhandle on January 7 and stopped at the North Palo Duro Ranch of Englishman James Cator. The cowboy recalled:

> We had just finished supper when his cattle began to bawl and make a fuss. "Another storm," remarked Cator. Sure enough it was another storm. Within an hour it was attended by a whirling, blinding mass of snow. I went at once to the barn to care for my horse. The visit might have cost me my life. I had the greatest difficulty, after being lost twice, in finding the house again, facing the wind and snow.
>
> For two days and three nights I was marooned at Cator's. On the third day I fought my way 15 miles north to our camp. Dead cattle were encountered here and there in the snow, across the fire swept district I covered. I reached camp expecting to find Bill dead. He was, however, alive and well. Our cattle had drifted off in the storm.
>
> Three days afterward I determined to ride to the drift fence on the breaks of the Canadian, where I had turned our cattle loose just after the fire. Twenty miles south I began to run into dead stock, then more and more lifeless animals were met until finally, when within a mile or two of the fence, I actually ran upon piles of them. Through these piles of poor, stricken brutes, cruelly sacrificed, I reined my horse. Never have I seen such a sickening, pitiful sight. I could

actually have walked for miles on dead animals, step-
ping from one to another. These were mostly natives,
belonging to northern ranges which had drifted across
the burned prairie. As long as they could travel, cattle
kept alive. Finally, the drift fence halted them. Here
they stopped, bellowed mournfully, bunched close
together as for a last protection, helplessly dropped in
their tracks and froze. Their carcasses lay where they
had fallen till spring. It was no use to ride further. I
finally reached shelter, spent the night and returned
to camp.[10]

Similar stories were told elsewhere.

Cattle walked off bluffs and fell to their deaths. Others got
stuck in bogs or in ditches. Some crossed frozen lakes but crashed
through the thin ice in the center of the lakes and drowned, froze
to death, or were trampled. Carcasses filled the Cimarron, Arkansas, and Canadian Rivers. Wolves attacked many of the helpless
animals. The deadliest trap, however, had been the drift fences
that stretched across the Panhandle.

The blizzard of 1886 became known as "The Big Die-up."

But more than cattle suffered.

LX Ranch cowboys John Arnot and Frank Ellis had been repairing fence near some burned grasslands and were on their way back
to headquarters when the blizzard hit. Soon, the wind and snow
forced them to seek shelter under a bluff at the Canadian River
breaks. They started a fire, but wet snow fell on the cottonwood
limbs and put out the fire. The two cowboys huddled together to
stay warm during the frigid afternoon.

Arnot recalled in 1944:

It was bright and clear the next morning. . . . We
had a time getting through the drift piled all round
our camp, had to ride the horses into it, back out and
gradually work it out. When we got through I told
Frank to ride in on the extra horse, get something to
eat and come back to me. It was about 12 miles to the
headquarters, but I had to circle around to miss the
drifts with the wagon. We got in after dark.[11]

25

Meanwhile, Les Carter had gone out to see if he could save some cattle when he discovered a covered wagon. A team of "bony sorrel horses" was dead in the harnesses with the reins leading from the dead animals through a hole in the wagon tarp. Carter looked inside the wagon and found an even ghastlier sight: the frozen bodies of a man, woman, and three small children.

John Peterson was driving an empty stage, pulled by a horse named Punkin and a mule named Kit, to Fort Elliott when the blizzard hit. Snow drifted into high banks, and the wind picked up sand, turning the storm into a "black blizzard." Blinded, Peterson drove the stagecoach off the road and into an old well some five feet in diameter. The driver walked back to the station on Commission Creek near Canadian, Texas, and found rancher Oliver Nelson and another man there. Peterson, Nelson, and the other man went back to the well, rescued Punkin and Kit, and returned to the warm comfort of the Commission Creek station.[12]

When the blizzard finally broke, Western cattlemen began to figure out their losses. An estimated fifty thousand dead cattle were found on the banks of the Arkansas River and along the railroad tracks near Dodge City, Kansas. Some one hundred and fifty thousand were thought to have died between Dodge City, Kansas, and Pueblo, Colorado. Losses in Colorado and New Mexico Territory were less, but the *Dallas Morning News* reported that "the storm has been the most disastrous to stock growing interests ever experienced in the Indian Territory."[13]

And in Texas?

James Cator's Diamond C herd was almost wiped out, but Cator recovered. Henry Cresswell's Bar CC Ranch also suffered greatly, but he, too, managed to bounce back. John Hollicot, manager of the LX Ranch, and his men skinned two hundred and fifty dead cattle per mile for thirty-five miles. That's eight thousand seven hundred and fifty dead cattle.

On the XIT Ranch, a herd that had been twenty thousand to twenty-three thousand now was a "vestige."[14] Total XIT losses, however, were reported at less than fifteen percent. Only eight hundred Bar C cattle were rounded up from its southern herd that had numbered eleven thousand. And the Quarter Circle Heart outfit in Donley County went out of business after reporting losses of about fifty percent.

J.H. Buchanan estimated losses along the drift fences at one hundred and fifty thousand. The *Drovers Journal*, however, expected the losses to reach two hundred thousand. Some reports stated that twenty-five percent of the drifting cattle from the northern ranges were dead but only one percent of the native cattle had died.

Not everyone bought that theory.

"It is not possible that on the same range and under the same conditions for the one to be wiped out and the other to lose only one percent," the *Texas Livestock Journal* reported.[15]

The *Livestock Journal* is undoubtedly right.

Some ranch managers, however, may have used the "Big Die-up" to their advantage, overstating the losses to bring the book count, the estimated number of cattle, closer to the head count, the actual number of cattle.

But even that can't diminish just how devastating the blizzard had been to the cattle industry, not only in Texas, but across the West.

Cowhides were shipped to market by train loads, and one merchant bought forty-five thousand hides alone. The Texas Cattle Raisers Association asked ranchers to send the hides to Dodge City so they could get a better estimate of the losses. More than four hundred thousand hides had arrived by June 1. Thousands more remained on the plains. Losses ranged from twenty percent to ninety percent among the larger ranches. Many small ranches went under.

Cattle prices had plummeted back in December, dropping from thirty dollars a head to ten dollars—if that. Now facing an uncertain future, ranchers realized they had to rethink the system of open range. The hard winter of 1886-87, which had less effect on Texas than it did on the northern plains, ended the era of open range. Instead of using barbed wire to build drift fences, ranchers began fencing in pastures and fields. Herds were reduced and/or improved. Drift fences were taken down. An 1887 state law made fences that crossed or enclosed public lands illegal.

The day of the drift fence was over in Texas. Cattlemen realized the fences didn't work after all. One Coke County rancher wrote:

> Drift fences were no account because cattle from the north would drift down in winter and stack up against these fences and die by the hundreds. All the grass north of the fence lines would be eaten so somebody's winter range would be destroyed.[16]

Winter ended, but the plight of ranchers (and farmers, too) worsened in 1886. The droughts that followed in the summer of 1886 and in the early 1890s would bankrupt and shatter the dreams of settlers.

## Chapter 5

# The Indianola Hurricane of 1886

### *"That town is a complete wreck"*

Indianola was jinxed.

More than a decade had passed since the terrible hurricane of 1875, and those who remained in the city were trying to reclaim Indianola's status as "The Mother of Western Texas." Mrs. Ernestine Weiss Faudie noted that "Capital was timid about investing again and most of those who lived [through the 1875 storm] moved away to escape another like fate."[1] Fewer than one thousand people resided in Indianola by 1880 (compared to the more than five thousand in 1875). Freight and passenger traffic had diminished, so Indianolans tried to bill their city as a coastal resort.

But Indianola was doomed.[2]

The year 1886 had been hot, dry, miserable, hardly the climate to bring in tourism dollars. The unprecedented drought garnered national attention. Crops failed. Livestock perished. Only one inch of rain fell in Weatherford during the months of April, May, and June combined. In August a citizens committee of farmers met in Ballinger, "seeking relief for the destitute caused by the extreme drouth and total failure of crops" in northeastern Runnels County. A similar appeal meeting was held in Anson later that month for Jones County. The average rainfall in Texas that year was 22.75 inches. In Corpus Christi, P. Doddridge had given one hundred barrels of water to the poor that summer. In Galveston, the drought seemed to bring out more mercenary attitudes; water sold for ten cents a bucket—compared to a nickel for a beer.

Rains had fallen in South Texas that summer. On June 14 a tropical storm hit the Louisiana-Texas border near the Sabine River. Winds reached fifty miles per hour, tossing telephone poles and washing away railroad tracks. Galveston residents watched the tide come within inches of the high mark set during the 1875 storm. Indianola would experience an even worse storm two months later.

August 18, 1886 was cloudy and humid in dry Indianola, with strong easterly winds filling the air with stinging clouds of dust and sand. Signal Service observer Issac A. Reed received a telegraph warning that the West India hurricane had moved into the Gulf of Mexico "and will probably cause gales on the coast of the eastern Gulf states tonight." No instructions came from Washington to raise the warning flag. By the time the signal came the next day to raise the warning, it was too late.[3]

Perhaps some didn't wait for the official word from the Signal Service. Faudie recalled that some "read the signs in time to evacuate the town and the havoc was not as destructive to the lives of the inhabitants." The 1886 storm's death toll was estimated at only between twenty and fifty, but the storm turned out to be quite destructive.[4]

The hurricane made landfall on the night of August 19. By morning, winds had reached seventy-two miles per hour. "Buildings which stoutly withstood the great cyclone of 1875, went down as if made of pasteboard," the *Victoria Advocate* reported August 28. Runge & Co.'s grocery, on Main Street's higher ground, toppled to the wind. Several people gathered for protection at the Signal Office across the street. Dr. H. Rosencranz; T.D. Woodward, chief clerk in the custom house; telegraph operator L.H. Woodworth; C.H. French; W.J. Morrison; and John S. Munn, who picked the wrong time to visit Indianola from Victoria, were among those at the Signal Office. Woodworth had been asked by Reed to work the telegraph, but by then the lines were down. As the pounding of the storm intensified, the building swayed. The office couldn't hold out much longer, so the occupants suggested they move to Tony Lagus's grocery store. Reed declined at first, saying he didn't want to abandon his post, "but at length decided to leave the tottering building." French and Munn left first, followed by Morrison,

Woodward, and Woodworth. Before leaving, Reed decided to reset the anemograph to get an accurate measurement of the wind's velocity. The decision turned out to be a fatal mistake for Reed.[5]

Rosencranz and Reed reached the sidewalk when the building collapsed, and the two were pinned by fallen timbers. "The water was several feet deep," the *Advocate* reported, "and it is supposed that both were drowned. The next day their bodies were found near where they had last been seen alive, Capt. Reed on the pavement, while Dr. Rosencranz lay a few feet out in the street." Accounts say the two were badly burned before they drowned; perhaps they were burned to death.[6]

Munn, an attorney, gave an overly dramatic account of Reed's last moments in the August 27 edition of the *Jackson County Progress*: "'Well,' [Reed] added, 'I will screw this down so that if the office goes and I am lost it may be found and read.' Then, turning to me said he, 'the water is up to my waist in my house; my family may be lost but my post of duty is here,' and heroically he stayed. The building cracked and I left, Capt. Reed saying he was coming. I never saw him again alive."[7]

The winds measured one hundred and two miles an hour when the Signal Office was destroyed.

On August 26, Reed's wife, Alice, wrote to her son, S.G. Reed, to tell him of his father's death:

> Your dear papa was drowned in a terrible cyclone at 6 o'clock on the morning of the 19th. The storm had been raging all night and as he was ever conscientious as to the discharge of his duty, he was at the office all night and had just taken his last observation and came down stairs with four others as the building was shaking very much. They were standing on the sidewalks in three feet of water, except Dr. Rosencraus [*sic*], who was standing in the doorway in front of the building, underneath the gallery. When someone screamed 'the house is falling, run,' three escaped, but your papa was a second too late. The building falling forward, he was caught under it. When they came down out of the office they left a lamp burning; it upset, caught fire and burned up two blocks."[8]

31

Fire had not played a part in the terrible 1875 storm, but the flames proved disastrous eleven years later.

"The fire began in the signal office; in a moment the flames flashed up and the cinders and smoke became almost suffocating," Munn recounted. "We stopped for a while and gazed with horror on this unexpected danger. It was then that the ladies shed tears, every lip faltered, every cheek blanched, how were it then possible to escape death. The most dreadful appeared inevitable as a lady said very [demurely], 'Well, I guess we have our choice, to be drowned or roasted.'"

Munn also noted that the bodies of Reed and Rosencranz had been badly burned: "Dr. Rosencrans' head was consumed, save a portion of his face, his arms were up over his head as though trying to protect from fallen timbers perhaps; his flesh was also burned from his hands, shoulders and back a large plank was clasped between his knees. Capt. Reed lay on his back, one knee raised, when it was burned nearly in two; his hands up and fingers burned off; his face was not disfigured."[9]

As the fire spread, refugees were forced to abandon the shelter and face the storm. "The spray cut the flesh like shot," Munn said. Fueled by the wind, the flames moved from the destroyed Signal Office, engulfing buildings on the east side of the street—August Frank's warehouse, Lagus' grocery, Steinbach's market, and Villeneuve's liquor store—before leaping to the west side, sending a hotel, bake shop, dry goods story, drug establishment, Antone Bauer's home, and other buildings in flames. By then, the water was four feet deep in the streets, and the spray somewhat protected the burning buildings. The lower part of Regan's store wasn't burned, not that it mattered. "This space is filled with a mass of debris from the fire, largely consisting of half-burned goods of every description belonging to the dry goods trade," the *Advocate* reported. "Regan's loss by the fire is thought to be about $20,000, upon which there was no insurance."[10]

Recalled Munn: "The howling blast, the roaring sea, the crash of falling timbers, the explosion of powder in stores, the crackling of flames as they shot up from, and lapped over the doomed buildings which rapidly yielded to wind, wave and fire, tottered, quivered and shrieking, fell into ruins and disappeared."[11]

One man, forced to weather out the storm south of Palacios Point on the Matagorda Peninsula with family and friends, recalled: "We saw a fire reflected in the sky in the direction of Indianola, we little dreamed the cause of it at the time, but later we knew it must have been the reflection of the burning city."[12]

Water and fire proved tragic.

"The water came up so suddenly and rapidly that it has been considered a tidal wave," Alice Reed wrote. "The water in our house was seven feet deep. . . . It was a most terrible time, people fighting against fire, water and wind—I have been almost paralyzed by this sudden blow. . . ."[13]

At the quarantine station, Dr. Fisher (or Fischer) and his wife and Captain M.S. Mahon (or Mahone) left the hospital building and fastened a yawl to some salt cedars with a long chain. They rode out the storm in the boat. Captain Adolph Steinbach and a woman named Martha Ellis may have been with the Fishers and Mahons, according to the *Galveston Daily News*. "For nine hours they clung to their frail support for life, the angry waves at times almost wrenching loose their hands, grasping with the energy of [despair] the swaying cedars." The quarantine station was washed away, and "nearly every house on the upper end of the island was swept away." A woman named Bettie Meade and two children drowned. Almost all livestock also perished, and the *News* reported that "the balance will probably die for want of water." The Fisher-Mahon party was rescued by Captain Sam Borge of the schooner *Tidal Wave*.[14]

"The scene of the saddest incident" happened at the residence of a Mrs. Shephard, where she took shelter along with her daughter, Mrs. Max Luther of Corpus Christi; two grandchildren; Mrs. Shephard's son, seventeen-year-old Henry Shephard; Mrs. Hodges of Cuero and two twins boys (their ages given as five or six); and Mrs. L.M. Crooker, Hodges' sister. By 4 A.M., with water knee deep in the house, Henry Shephard took the party into Mrs. Shephard's room in another building. The nine people were swept into the melee when the outbuilding was "wrenched from its foundation" by the water and wind. The roof trapped Henry underneath the water, but he managed to free himself and found he was holding onto his sister. Henry pulled Mrs. Luther onto the floating roof.

Luther had been holding onto her infant, "but in the fierce struggle under water the child had been swept away and lost." Henry and his sister somehow managed to stay on the roof, despite being "Surrounded by masses of shattered houses, the water, driven by the howling wind, boiling and foaming around them. . . ." While his sister lay stricken with grief and fear, Henry caught the attention of "a young man named Smith," who brought brother and sister back to Indianola.[15]

The rest of the party wasn't as fortunate. Mrs. Shephard, Mrs. Crooker, Mrs. Hodges, and one of the twins were believed to have been killed by falling timbers. Their bodies were found by the end of the weekend, and Hodges, Crooker, and the two Hodges children were buried in Cuero on August 24.

John Smith found Mrs. Crooker's body at cattle pens three and a half miles from town. She had her gold watch and chain, several diamond and gold rings, a diamond broach, bracelets, "ear appendages," and a purse containing a twenty-dollar gold piece and ten-dollar bill. "Dr. Hodges and Mrs. Crooker were said to have had a premonition of danger in going to Indianola and the latter went reluctantly. The party, however, especially the twin boys, except Mrs. Crooker, were apparently in fine spirits all the way, the latter seemed depressed and anxious. They arrived in the doomed city in time to meet a terrible death."[16]

Some had better luck. The *Advocate* reported that Mrs. E. Westhoff and her infant child, along with six other people had been "carried across the bayou clinging to pieces of wreck. The infant child of Mrs. Westhoff was three times beneath the waves but each time rescued by the heroic mother." All eight people survived.

The *Advocate* also recounted one humorous tale: "A man drifted nearly 10 miles on a board, on one end of which was a large rattlesnake. The man and snake were saved."[17]

Destruction wasn't limited to Indianola, of course.

"WRECKED BY THE WATERS," the *Galveston Daily News* headline read Saturday, August 21. The storm struck that city on Thursday and "raged violently." Streets were flooded by "the highest tide for many years," four or five inches more than during the 1875 storm, and many residents took refuge at the University of St. Mary. Father O'Connor, the university president, said the refugees

"were all badly frightened, and two or three were quite sick."
O'Connor went on, telling the *News*:

> They came away from their homes hurriedly, and
> those who brought anything with them brought only
> small articles which they were able to save during the
> excitement and hurry of their departures. The sight
> this morning was a novel and curious one, and at the
> same time very sad and touching. This lower floor was
> devoted entirely to the colored people, and every
> nook and corner was filled by them. The second floor,
> the school room and dormitory were given up to the
> white people. There was not a spare foot of room any-
> where in the building."

Although railroad traffic had been stalled and the night had
been "truly hazardous to a large portion of Galveston's citizens,"
the *News* reported only three fatalities.

At Shoal Point, "Our beach is strewn with debris from one end
to the other, consisting of chickens, dogs, doors, sashes, window
frames, tubs, buckets, barrels, and in fact all kinds of stuff, and several parts of boats of most all sizes." Crops were destroyed, but
"Everybody is cheerful. All have gone to work repairing damages to
fences. . . ."

At Goliad, the *News* reported, "Many houses were blown down
. . . Yet no loss of life, so far, has occurred. . . ." Six or seven houses,
the Temperance Hall, and a cistern factory were blown down in
Rockport. Cuero, Houston, Gonzales, Sequin, Luling, Lockhart,
New Braunfels, Weimar—even as far inland as Abilene—also suffered damages. At San Antonio, the barometer hit 29.03, winds
were estimated between seventy and eighty miles per hour, and
damages reached $250,000. Yet the storm also had bright spots.
Although winds had been estimated at seventy-five miles per hour
at Corpus Christi and the bay water receded, leaving boats at
anchor "high and dry for two hours," six inches of rain fell—"a glo-
rious rain at last"—ending the drought. Rains from the storm also
eased the dry conditions elsewhere in South and West Texas.[18]

Indianola, however, had no bright spot.

"A train has just arrived from Indianola bearing the sad news of another terrible disaster there," the *News* reported August 21. "The water is said to have inundated the town worse than during the great storm of 1875. The railway track is washed away for two and a half miles this side of Indianola, rendering communication extremely difficult."

News soon spread just how complete the destruction had been.

"There is perhaps not a half dozen houses, including the court house and jail, that are tenable, being twisted, unroofed or totally wrecked," the *Advocate* said. "The prairies for miles is strewn with wrecks of boats and buildings, merchandise, furniture and dead animals."[19]

The "oyster shop" of John Mathuly, which had survived the 1875 hurricane, had no such luck this time. Five churches were pounded, but the Catholic church survived. The Morgan wharf was washed away as was the bridge across the bayou, trees on the square had been reduced to "charred trunks," and salt water contaminated underground cisterns. At least seven boats in the bayou had been wrecked, and many crews were missing.

"That town is a complete wreck," a correspondent reported in the August 24 *Galveston Daily News*, "not more than three or four houses having escaped the destructive force of the water."

Wreckage of the bridge over Powderhorn Bayou
after the 1886 Indianola hurricane.
*Texas State Library & Archives Commission*

Relief parties once again were organized. One of the first to arrive was a Dr. Reuss of Cuero. "As soon as he learned of the disaster he left for the calamitous scene, and exerted himself in every possible way to relieve, as much as possible the sufferings," the *News* reported. "From Friday evening until Sunday afternoon he remained at his self-imposed post of duty, and did much to lighten the burdens pressing heavily upon the worn and heart-weary people who surrounded him."

Many survivors had been left with only the clothes they wore. The *Advocate* reporter commented: "It was pitiful to see men, women and children with pale and anxious [countenances] eagerly seeking information of missing relatives."[20]

For most, they had had enough of Indianola. The quarantine station was rebuilt at Port O'Connor, and the county seat was soon moved to Port Lavaca. The myth, however, is that Indianola vanished immediately after the August 1886 hurricane. It didn't, of course. Some people remained, but the city was jinxed.

A second storm pounded the Texas coast in late September, this one dropping 25.98 inches of rain in Brownsville. Nor was what remained of Indianola spared. Water from Matagorda Bay flooded the streets again, reaching waist high, and winds hit sixty miles an hour. Enough was enough. Most people fled the doomed city by 6 P.M. And yet another storm struck Texas in October, and this one even deadlier than the August storm. Mercifully, Indianola was spared this time, but Sabine Pass was "virtually swept out of existence," the Catholic church in Orange was destroyed, a storm surge of seven feet flooded Johnson's Bayou and Sabine Pass, where "nearly every house in the vicinity was moved from its foundation." A woman reportedly crossed Sabine Lake on a feather mattress. Death figures given for the October storm include one hundred twenty-six, one hundred fifty, and between one hundred seventy-five and two hundred. So the last storm of the 1886 hurricane season to hit Texas was far deadlier than the August storm, but that's probably because Indianola had not recovered to its 1875 heyday.

Indianola died. Buildings—what was left of them, anyway—were torn down to be rebuilt in Victoria, Cuero, or elsewhere. On November 10 the Commissioners Court held its final meeting in

Indianola. An April 1887 fire consumed most of what remained of the town, and the post office ceased in October of that year.

The question that cannot be answered remains: Would Indianola have survived had the city been relocated to so-called safer ground, as had been planned after the 1875 storm? Would she have rebounded to remain Galveston's rival as a shipping port? Today, a state historical marker and a statue of Robert Cavelier, who landed near here in 1685, a few buildings, and foundations mark "The Mother of Western Texas" at the end of state Highway 316.

"Of the many ghost towns in Texas," one writer noted, "none died as tragic a death as Indianola."[21]

Once a bustling port city that rivaled Galveston, Indianola has been reduced to memories and historical markers after killer hurricanes in 1875 and 1886.
*Author's Photo*

## Chapter 6

# *The Drought of 1890-1894*

## *"The air was rancid"*

The 1890s drought stood out in people's memories, despite its lesser severity than other droughts from the last few centuries.[1]

Well, it was bad enough.

In South Texas, they called it the "Great Die."

Scores of livestock perished, and the bottom dropped out of the cattle market. In West Texas, forty-seven counties reported drought in 1890-1894. Robert J. Kleberg II, manager of the great King Ranch in South Texas, described "The cattle wandering over the prairie perishing, how their lowing wrung my heart." The ranch's Mexican vaqueros known as *Los Kineños*, or the King's Men, skinned dead cattle to salvage the hides. One *Kineño* said he skinned more than seven hundred cattle and pulled more than two thousand out of a muddy Caesar Creek: "They would come to the almost dried up holes there and get stuck in the mud and often stay there until they died," he said.[2]

Crops succumbed to the heat, and parts of Texas were plagued by wildfires, sand and dust storms, and invading grasshoppers. Fort Davis suffered through a three-day dust storm in November. Prairie dogs died of thirst, and the bleached bones of cattle littered the land. In Castro County, "the air was rancid" because of the number of dead cattle.[3]

Droughts weren't new to Texas—and still aren't. Spanish explorer Alvar Núñez Cabeza de Vaca discovered a village near Presidio where it hadn't rained in two years, and the natives, thinking him a god, pleaded with him to make it rain. In 1756 the San Gabriel River dried up, forcing missionaries and Indians to abandon their settlement. And the drought of 1886 left many Texans on relief.

What many non-Texans find incomprehensible is just how long these dry conditions last. Noted novelist Elmer Kelton illustrates that point in the introduction to his classic novel of the 1950s drought, *The Time It Never Rained*, recalling when he had asked a lamb buyer about the conditions in Kansas:

> He said they were tough; he had suffered through three drouths that summer. I did not totally relate to his terminology. I told him we were already in the third summer of the same drouth. As it turned out, we had three more to go.[4]

The year 1892 had been a wet year in North Texas, but the following year was dry. The drought continued through the fall and early winter, and when it did rain, the weather was violent. Albany, Dallas, and Paris suffered damage during an April 11 storm, and an April 28 tornado left more than twenty people dead in Cisco.

In Nueces County, where the mean annual precipitation is 30.2 inches, Corpus Christi recorded 25.6 inches in 1891, 20.6 in 1892, and 20.5 in 1893. In Bexar County (29.1 inches mean annual precipitation), only 18.2 inches fell in 1893. It was just as bad elsewhere. Precipitation was 4 percent below normal in the spring of 1891, 14 percent below normal that summer, and 38 percent below normal that fall. The next year, precipitation was 21 percent below normal, and it was 33 percent below normal in 1893. The state mean of .03 inches in October 1893 is among the lowest ever recorded for that month.

Those are just numbers. What they don't tell is how farming and ranching in Texas flourished in 1890. Rains had ended the drought of the 1880s. Abilene's cotton crop brought in an estimated $300,000 in 1890. Land prices rose astronomically in West Texas, where several new counties were being organized. Things looked bright.

And then:

Fort Stockton reported that "the springs are lower than they have been in years," while Big Spring noted in March that "there has been no addition to the surface water since October." Water was scarce in many Texas communities in 1890, and it quickly got worse.[5]

Hidalgo County lost 7,498 of 28,404 sheep, the U.S. Census of Agriculture reported in 1890. Starr County lost 9,033 of 59,999 sheep. In 1893 near San Diego in South Texas, rancher Jorge Alaniz gave away eight hundred head of cattle because he couldn't feed them.

Adding to the weather catastrophe was a pitiful market for cattle. Range cattle were worth $5 a head in 1892. In the Lower Valley in 1891-92, you could buy cattle for $4 a head. In December 1893, H.M. Field sold forty-eight head at $8 each; more than a year later, another sale averaged less than $6 a head. Compare these prices to the $24 to $40 a head that cattle sold for during the heyday of the trail drives to Kansas in the 1870s. The Panic of 1893 certainly didn't help matters. The sheep and wool market also suffered. South Texas sheep sold for $1 to $1.25 a head—and sometimes less—in 1893, after selling for $3 in 1878. Partly because of an increase in foreign producers, the price of wool had dropped from twenty cents a pound in 1880 to seven cents in 1893.

At the Matador Ranch, founded in 1878 near the rim of the Caprock in West Texas, the drought reduced the 1893 calf crop by ten thousand head. In Borden County at the MK Ranch, which branded six thousand calves in 1893, only one hundred and sixty were branded the following year. The 439,972-acre Spur Ranch of West Texas, which earned $109,800 in cattle sales in 1892, dropped to $16,826 in 1893.

"Crops are hurting, cattle are suffering and rain is needed," the *Dallas Morning News* reported in 1892.[6]

Looking for help, a rancher wrote to "Hester's Weather Forecasts," published in newspapers, in 1892, asking for insight into the weather in his region. Hester responded that his service was available only to subscribers. The rancher sent in the $3 fee, along with the latitude, longitude, and altitude of his ranch. Hester replied that "the horoscope shows no relief from present conditions for some time to come," and he was right.[7]

The Matador Ranch was forced to buy and lease hundreds of thousands of acres to contend with the lack of grass, obtaining 214,000 acres from the XIT and leasing more land near White

Deer. Trainloads of cattle were shipped elsewhere. Even King Ranch needed help. In 1892 the drought forced Kleberg to send twelve thousand head of cattle north to the Indian Territory of present-day Oklahoma. Many West Texas ranches also had to send starving cattle to the Indian Territory.

That year, Panhandle cattleman Charles Goodnight had been asked by the *St. Louis Globe-Democrat* if a farmer could make a living in the country. Goodnight replied:

> Yes. But he can't make money. He may, by hard work, do a little better some seasons than a living, but he can't get rich. The only way a farmer can do well here is to combine stock-raising with farming.[8]

Ranches and farms failed. As historian T.R. Fehrenbach notes:

> Only a few succeeded. Year after year, stockmen, or farmers turned farmer-stockman, bought the failures out, in a process that went on for many years, and was accelerated with every recurrent, hideous drouth, from 1887 to 1917, and from 1918 to 1933.[9]

The twenty-five-section MK Ranch died. Banks closed. Towns died. Those who survived were blessed with capital, luck, and foresight.

Windmills, first manufactured in South Coventry, Connecticut, in 1854, were commonplace on the plains by the 1890s. Sorghum and man-made water holes helped ease the drought on the Spur Ranch. The massive XIT Ranch in the Panhandle experimented with drought-resistant plants and had begun "farming" with alfalfa, oats, millet, sorghum, and Indian corn—plus five thousand trees had been planted in 1886. The XIT also spent $500,000, building three hundred thirty-five windmills and one hundred dams, digging water tanks, and spreading sacks of stock salt on the bottom of the new tanks.

Perhaps the most forward-thinking operation was King Ranch.

Robert J. Kleberg II was more than just the ranch manager. He started out as one of ranch founder Richard King's attorneys in the early 1880s. King died on April 14, 1885, and his widow, Henrietta, appointed Kleberg full-time business manager on

January 1, 1886. On June 17 of that year, Kleberg married the Kings's youngest daughter, Alice.

In an area that had been plagued by primitive agricultural methods, Kleberg had helped turn King Ranch, which had more than six hundred thousand acres, into a more efficient operation, building cross fences, improving breeding stock, and clearing land of mesquite. Kleberg understood that the size of the ranch would help during the drought—the cowboys could move the cattle to various pastures—but he still needed water.

"Where I have grass, I have no water," Richard King had complained. "And where I have water, I have no grass."[10]

Cattle on the ranch would eat the prickly pear cactus—including the spines—to get the water, but it proved too painful for them to eat enough, and many died. Later, the needles were burned off with gasoline torches. "So, when we began to burn off the needles," Kleberg said, "the cactus pad provided a measure [of] insurance against drought. One man in a day could burn enough to save 300 cattle."[11]

Kleberg experimented with new and old ways of getting more surface water. Expanding the water tanks and digging shallow wells proved to be a failure. He even took part in a U.S. Department of Agriculture experiment aimed at making rain by sending balloons filled with explosives into the atmosphere (see next chapter). Finally, the first artesian well was brought in during the summer of 1899.

Kleberg described his emotions when a well at Leoncitos pasture finally came in and he saw "cattle lapping the water from the ground."[12]

Said Kleberg:

It flowed 200 gallons a minute. Every drop meant life. I just went over to a tree and cried. It meant a lot. It meant that we would hear the whistle of the locomotive. It meant cotton and forage crops, and a cotton mill, and a town, and good schools. They have all come. We have had the railroad since 1904. We have 10,000 acres planted in cotton and forage, and 2,000 acres in Rhodes grass. The cotton mill offers steady employment. We made a town at Kingsville. There are four schools on the place; our Mexican children are getting an education. Large parcels of fertile land have been cut into farms and sold on easy terms. The country is settling. Progress has come to us down here, and we have gone out to meet it.[13]

The drought eventually broke, but the rest of the 1890s remained dry.

In his 1931 book *The Great Plains*, Walter Prescott Webb pointed out that government officials "have warned the people over and over for nearly half a century that there is no basis for the belief that climate in any place is subject to appreciable change, either in temperature or rainfall."[14]

Indeed, the Chief of the Weather Bureau reported in 1896:

It was clearly shown from the investigation made that periods of alternating wet and dry weather were characteristics of the seasons forty and fifty years ago, and that there was no general law governing the recurrence of years of drought or abundant rainfall.[15]

## Chapter 7

# The Rainmaking Experiment of 1891

## *"Producing rainfall by concussion"*

 King Ranch livestock manager Sam Ragland once said, "When you can see the cow chips floating, then we've had a good rain."[1]

Rain. Ranchers and farmers pray for it, especially west of the 98th Meridian. There's a wonderful scene in the movie *Dancer, Texas*, where one West Texas rancher greets another, and the two of them simply stare at the sky without talking. A Hollywood over-exaggeration? Not hardly.

In the Wild Horse Desert of South Texas, water meant life. Consider these excerpts from letters Richard King wrote to wife Henrietta in 1883:

> "Grass is very short . . . still some cattle dying here not enough of rain."—March 18
>
> "Everything lovely at home but grass and that is growing short—it looks to night like rain hope it will rain . . . hope and pray for a good rain."—March 24
>
> "We all miss you at home we are all well but no rain so far and it makes us all feel bad stock is suffering for grass but we hope for the best as we have at all times have our portion of good luck. . . ."—June 22
>
> "Have commenced shearing but have stopped on account of rain as it has rained every day since my arrival . . . the finest grass I have seen for 20 years at this place everything full of water . . . the only thing that is missed here is Mama and my pets. . . ."
> —June 27

"We have had fine rains and plenty of them here the stock is all right I think now the grass is growing fast ... it does look more cheerful now since the rain and in six or eight days we will have fine grass here and all around us in fact it has been a good and general rain in this section we are doing the best possible in matters but a great deal of bother—in matters here during the dry time every body was crusty and mad. . . ."—July 6

"I think we are all well and the grass in the yard is green once more thanks be to God for it—we were getting in a fearful fix if it did not have rain. . . ." —July 8[2]

Western frontiersmen, it seems, would try anything—would believe anything—to make it rain. Theories abounded: If you plowed up the land, it would hold moisture, thus increase evaporation, thus make more rainfall possible. Growing crops or burning the prairie would also cause precipitation. One theory required building a tall chimney over water and starting fires at the water's edge. The flames would be drawn over the water to the chimney and carry moisture into the air.[3]

So it should come as no surprise that King Ranch manager Robert J. Kleberg II took an interest in the U.S. Department of Agriculture's rainmaking experiment during the drought of the early 1890s.

The government-backed plan was the brainchild of Robert G. Dyrenforth, special agent of the Department of Agriculture, and Illinois Senator Charles B. Farwell. *War and the Weather*, a book by Edward Powers, included data that even in dry regions, "copious" rain fell after battles during which there had been cannon fire. In 1880 General Daniel Ruggles of Virginia patented his production of rainfall by causing explosives in the air.

More theories abounded:

The concussion from the explosions jarred the air, smoke caused a reaction of "nuclei or mechanical retaining points," the atmospheric pressure reacted to the concussion, the buoyancy of the produced gases and the heat to force a current up, causing a disturbance, the explosions generated electricity and friction, "producing polarization of the earth and sky ... inducing ... other conditions necessary for storm formation, electrical manifestation being a constant forerunner and concomitant of storms. . . ."

There were other theories, too, including several—understandably—that Dyrenforth "was quite unable to understand." Still, Dyrenforth "found innumerable instances where heavy cannonading had been followed by copious fall of rain, though there was nothing definite to indicate that there might not have been rainfall in each case without the firing." It seemed enough to warrant further investigation, so on February 27, 1891, Dyrenforth was appointed special agent to conduct experiments regarding the "idea of producing rainfall by concussion."

Dyrenforth had been educated at the University of Heidelberg and Columbian University (now George Washington University), served in the Union army—where he was wounded twice—during the Civil War, was a Austro-Prussian War correspondent, an attorney, and interim commissioner for the Patent Office in Washington.[4]

Dyrenforth, who had learned to work with explosives in the military, obtained balloons, bombs, and other equipment and began preliminary tests near Washington, D.C. The rainmaking crew included assistant John T. Ellis of Ohio's Oberlin College, aeronautical mechanic George E. Casler (King Ranch historian Tom Lea spells the last name Castier), and a balloon maker. Explosive-filled balloons took to the air. The noise wasn't met with total appreciation.

SMITHSONIAN INSTITUTION,
Washington, D.C., June 23, 1891.

SIR: Permit me to enter a protest against a continuance of the experiments in firing the balloons on the property adjoining my farm and residence. I have a herd of very fine Jersey cows, some with calf, and the tremendous explosion yesterday right over my barn was calculated to cause abortion. I have had this happen from a thunder storm and your bomb was worse than any thunder. It shook the house and alarmed my family.

Please, Mr. Secretary, move your dynamiters away
from "Oakmont," Piney Branch.
Yours, respectfully,
William J. Rhees,
Chief Clerk, Smithsonian Institution.

Rainy, humid Washington wasn't a good place for the tests anyway, but Texas was. Nelson Morris invited Dyrenforth and his crew to Morris's C Ranch near Midland, agreeing to pay for local expenses. Dyrenforth accepted, and the rainmaking crew reached the ranch on August 5.

"The region being entirely without stream or natural pool, every ranch is supplied with numerous wells from which water is pumped into surface tanks or ponds by wind-wheel pumps," Dyrenforth said. "Here and there are extensive depressions, called by the herders 'draws,' which, being dry, showed white alkaline efflorescence." It seemed the perfect place to begin the experiments.

Sixty mortar-like guns were constructed, and dynamite and rackarock were placed in prairie dog and badger holes. Kites, with dynamite sticks suspended by electrical fuse, were flown, their strings attached to mesquite, chaparral, and catsclaw, and balloons were filled with hydrogen. The explosive show began August 9. The next morning, it rained for two hours, "causing water to run into the 'draws,' and the plains to be drenched." Dyrenforth hadn't been prepared for rain so soon, so he could only guess at the amount, estimating an inch had fallen. Any celebration, however, was short-lived when the experimenters learned that a heavy rain had also fallen northeast in Abilene.

Experiments resumed two days later, but no rain fell. The firing had been light that day, as it was the following three days to save explosives and wait for better conditions. A few heavy explosions were sent airborne later, but rain, if any, was slight. On August 25 heavy firing began once more. Balloons rose and exploded, and ground explosions began at sundown, fired at short intervals. The bombardment stopped at 11 P.M.

"At about 3 o'clock on the following morning, August 26, I was awakened by violent thunder, which was accompanied by vivid lightning, and a heavy rainstorm was seen to the north," Dyrenforth reported. It fell heavily for hours in the distance, but only slightly at the ranch house. "While the thin edge of the cloud was overhead, a few charges of dynamite were fired near the ranch house. A few moments after the first explosion the first overhead rain began, and after each subsequent explosion rain could be seen falling from the clouds overhead in what appeared to be a heavy shower, but the air was still so dry that at first no rain and afterward only a sprinkle reached the ground. After each explosion the quantity of rainfall increased."

No rain was reported to the west in New Mexico, at Fort Stanton and Santa Fe, or at El Paso, San Antonio, or Abilene. "The storm seems to have been entirely local," Dyrenforth said.

Positive newspaper reports, from the *Rocky Mountain News* to the *Washington Post* and the *Chicago Times* and *Tribune* to the *New York Sun*, had the crew beaming.[5]

Dyrenforth left for Washington, putting Ellis and Casler in charge of the experiment. The crew took off for El Paso, arriving in September where Lieutenant S. Allen Dyer of the Twenty-third Infantry, stationed at Fort Bliss, was ordered to command a detail to assist with the project.

Heavy bombardment began on September 18. It didn't rain in El Paso, but Ellis reported that "A heavy dew had fallen during the night, an occurrence which, I am reliably informed, had never been known before in that region." However, reports later informed Ellis that "soon after midnight rain had begun to fall *within a few miles of El Paso*, to the south and southeast, evidently coming from the clouds *which had formed over the city*, during the explosions, and, between midnight and morning, a heavy rainstorm had passed down the Rio Grande valley, copiously wetting the valley, including a few miles of the contiguous portions of Texas and Mexico.

The El Paso experiment was cut short. Lieutenant Dyer thought that the explosions should have continued longer, "But the material furnished by El Paso was so small in quantity that it was necessary to stop the work before the desired results had been affected."

By then, however, Robert Kleberg had taken an interest in the experiment. "Everything ordered," he wired Ellis on September 12. "Come to Corpus Christi." Kleberg and Henrietta King were willing to help pay for some of the supplies, forking over an estimated $900 in explosives and equipment and $746 in traveling expenses. The cost of government stores totaled $633.52.

The rainmaking crew set up station at San Diego, about twenty-seven miles northwest of King Ranch's Santa Gertrudis headquarters. At 1 A.M. October 16, with the Weather Bureau reporting an improbable chance of rain, the experiment began. Dynamite and rackarock were set off, and balloons were exploded at 2 A.M., 4 A.M., and 7 A.M. Three-pound charges of dynamite were blasted at five-minute intervals until 12:30 P.M. A few clouds gathered but disappeared by nightfall. At 5 P.M. Ellis and Dyer renewed the explosions, continuing to fire dynamite until 8 P.M. A ten-foot balloon was exploded at 6 P.M. and another at 11 P.M. Three more were sent up and blown up at seventy-five-minute intervals, the last being destroyed at 3 A.M. The sky remained clear.

The bombardment resumed on the evening of October 17. At 9:45 P.M. the first balloon went up. Ellis recalled: "As it ascended, easily visible by the fact that it was a moonlit night, the men at the batteries made ready, and the moment the flash of the explosion seen, the dynamite and mortar batteries and the cannon were exploded, the flash of the balloon serving as a signal for these explosions, the roar of the ground explosions mingling with the crash of the explosion of about 1,000 cubic feet of oxy-hydrogen gas produced a concussive effect of tremendous force, and the ground was shaken for miles, as we afterwards learned, in every direction."

Explosions continued at a rate of ten a minute until 11:30 P.M., while hundreds of people from San Diego and other towns "crowded together in a frightened mass" until assured that it was safe to go home. The cannonade continued. At 4 A.M. the last balloon exploded, "and rain began to fall at that point immediately after the explosion." More bombs and rackarock charges were quickly set off, and scattered drops fell. "In five minutes a steady rain was falling and in another ten minutes, the water was pouring down in torrents," Ellis reported. Almost a half inch fell before 5

A.M. "Fifteen minutes after the rain ceased, the northern half of the sky was entirely clear, and in an hour there was hardly a cloud in sight. Our weary men 'turned in' after seventy-two hours of work with only five or six of sleep." On October 19 the crew returned to Corpus Christi and disbanded.

A success?

Dyer reported that it had rained in South-Central Texas, "being central and heaviest at the point of our operations and directly north of that point, where the effect of the explosions would be the greatest. The storm was local in character and had come unforetold and unexpected by any office of the Weather Bureau. There was certainly every possible appearance of the rain having been the result of the operations at San Diego."

Dyer certainly believed in the project, and he said everyone present thought the same. "One gentleman informed me that this rain, comparatively light as it was, had been worth to him alone a number of thousand dollars," Dyer wrote, "and that a large area of country in that vicinity had been greatly benefitted by it."

Others agreed:

"It affords me the greatest pleasure to say that I am satisfied with the result," Duval County rancher H.J. Delamer wrote. "From what I saw and from what I heard from those who watched throughout the night, I am of the opinion that the rain which fell on the morning of the 18th, was produced by the explosions."

Wrote Dr. Lincoln B. Wright of San Diego: "I feel safe in saying that you evidently produced rain, or, to be more explicit, that rain would not have fallen without the effects produced by your experiments."

Judge James O. Luby of Duval County was also satisfied. "The previous desultory explosions unaccompanied or followed by copious showers caused many to be skeptical of the final results; your choosing the night of the 17th, on a rising barometer and an expectant dry norther, made me think you courted failure in the face of a frowning world.... What was my surprise, after retiring for the night, to hear the patter on the shingles...."

G.W. Fulton of Gregory, Texas, who was present at the El Paso experiments and gave $300 to the San Diego cause, wrote: "I am forced to the conclusion that the rainfall was the direct result of the explosions. . . ."

Robert Kleberg, however, wasn't totally convinced.

The explosions woke him up at Santa Gertrudis at 11 P.M. Outside, he found the sky clear and moon and stars bright. Kleberg wrote:

> I retired about 2 o'clock A.M., and at 4 A.M. I was again awakened by the rain falling on the roof of the house I was occupying. At 7, when I arose, but few clouds were visible and a dry 'norther' or north wind was blowing. The rainfall at this place was very light, but a neighbor living 20 miles to the west of here and south of San Diego told me that on his ranch the best rain fell that night which had fallen there in three years. Now, I will further state in this connection, that, as all the 'northers' we have had this winter, before that time and since, were 'dry northers,' I do not think that the clouds from which the rain fell that night or morning, were brought on by the 'norther,' but on the contrary were driven off by the same, and it seemed to me that the rain which fell could fairly be claimed to have been produced by your experiment.

But:

"I do not think, however, that these experiments, which were made under the most adverse circumstances, have demonstrated, fairly and sufficiently, the theory whether or not rainfall can be produced by heavy concussions."

Kleberg said the Midland-El Paso-San Diego tests seemed encouraging enough for the government to budget more money for further tests. "There certainly is no subject of greater interest to our entire people than this, for in almost all portions of the Union has there been suffering from drought this year."

Others agreed, and not only Texans. Dr. William Taylor, correspondent of the British Museum and Smithsonian Institution, firmly believed that the explosions produced the rainfall at San Diego. "I consider the experiments highly successful, and sincerely hope that further experiments may be tried in this region."

Dyrenforth deduced:

First. That when a moist cloud is present, which, if undisturbed, would pass away without precipitating its moisture, the jarring of the cloud by concussions will cause the particles of moisture in suspension to agglomerate and fall in greater or less quantity, according to the degree of moistness of the air in and beneath the cloud.

Second. That by taking advantage of those periods which frequently occur in droughts, and in most if not in all sections of the United States where precipitation is insufficient for vegetation, and during which atmospheric conditions must favor rainfall, without there being actual rain, precipitation may be caused by concussions.

Third. That under the most unfavorable conditions for precipitation, conditions which need never be taken in operations to produce rain, storm conditions may be generated and rain be induced, there being, however, a wasteful expenditure of both time and material in overcoming unfavorable conditions.

The project cost about $17,000, of which the government spent about $9,000. The experiments, however, stopped in 1892. Officials, it seems, just couldn't decide "whether the rain was the result of the experiments or natural meteorological conditions."[6]

## Chapter 8

# The Brazos River Flood of 1899

---

### *"Now everything is gone"*

**A** hydrologist named Robert E. Horton once said: "A small stream cannot produce a major Mississippi River flood for much the same reason that an ordinary barnyard fowl cannot lay an egg a yard in diameter."[1] True enough, but Texas creeks, streams, and rivers can and have produced destructive floods, perhaps not as awesome and shocking as the Mississippi floods of the early 1990s, but horrible enough for Texans who live in flood areas.

The Brazos River flood of 1899 is one such example. The river had flooded many times before, including 1852, 1876, 1885, and many times since, but the 1899 inundation remains one of the most severe.[2]

The longest river in Texas, at eight hundred forty miles, the Brazos has a drainage area of 41,700 square miles. Settlements quickly popped up on or near the banks: Graham, Brazos, Granbury, Waco, Hearne. . . .

Located in southwestern Robertson County, Hearne was chartered on April 11, 1871, but the settlement goes back much further. The land, originally granted to Francisco Ruiz in 1830, was site of a tavern and stagecoach station in the 1840s, and in the 1850s cotton planters Horatio Ransome Hearne and Ebenezer Hearne bought ten thousand acres. The area grew, and with the arrival of two railroad lines to go with two major highways, the town was established in 1868. By 1885 the town had four churches, two cotton gins, two hotels, a Masonic hall, and a newspaper. In 1891 the Hearne Building and Loan Association was organized, only the sixth such institution in Texas. The population reached thirteen hundred in 1885 and would be at two thousand one hundred twenty-nine in 1900.

Hearne wasn't the only town growing along the Brazos in the 1890s. Fairchilds, in southeastern Fort Bend County, had attracted German settlers in the 1890s. In 1896 fifty northern Mennonite families, who practiced the Anabaptist religion, bought a league of land that included Fairchilds.

On June 28, 1899, a tropical disturbance made landfall near the mouth of the Brazos and slowly moved inland, dropping an extraordinary amount of rain mostly along the Brazos River basin from Granbury and Waco and on down to the Gulf of Mexico. Rainfall averaged 8.9 inches over sixty-six thousand square miles. Up to twenty inches fell from Temple to Palestine and between Victoria and Houston. In Hearne, a three-day deluge overflowed the rain gauge at twenty-three or twenty-four inches. An estimated 31.40 inches of rain fell on Hearne. In Franklin, more than forty inches were said to have fallen. At Richmond, where flood damages on the Brazos begin with discharges of about eighty-seven thousand cubic feet per second, the peak discharge reached two hundred thousand cubic feet per second. In Waco, two flood crests, on June 30 and July 1, were reported by Weather Bureau section director Isaac M. Cline of Galveston, with the maximum estimated at 34.4 feet. Waters rose over other flood gauges along the Brazos, and the San Saba, Neches, and Buffalo Bayou were also flooded.

Mennonite H.E. Unruh wrote of the weather leading up to the flood for the *Mennonitische Rundschau* on June 29:

> We read frequently about the weather in the "Rundschau" and since I have no other news I will talk about the weather here. Between February and June 24 there was only one good soaking rain. Otherwise the weather was pleasant with the exception of a few days when the temperature reached 98 degrees. The corn that was planted early was affected by the dry condition. Cotton which stands the heat better looks promising. Saturday, June 24, we had a drenching rain. . . .[3]

In Hearne, floodwaters rushed into town and continued to rise. More water came into the town by Sandy Creek, Lost Creek, and

Pinoak Creek and finally washed away the Houston and Texas Central rails and the private railway to Valley Junction. People couldn't leave town by wagon or buggy, and riding out on horseback was dangerous. Residents ran short of food. All along the Brazos, from McLennan County to the Gulf of Mexico, waters covered the basin two to twelve feet deep, and some reports said the Brazos was more than twelve miles wide in places. Cotton, corn, millet, peas, sorghum, sugar cane, and other crops were washed away. So were small tenant houses.

As H.E. Unruh continued in his letter to the *Mennonitische Rundschau*:

> Monday evening, the 26th, it started to rain and continued for 40 hours without an interruption. I have never before seen the creeks around here rise like that. Close-by two bridges were torn away by the water. . . .[4]

Residents found Richmond under water four to five feet deep. The Railroad Street bridge was also submerged, and the town became surrounded by water that reached into the prairie and in places was several miles wide.

Edmond Carter, who lived four miles outside of Richmond, and some of his neighbors found themselves cut off by water six to twenty feet deep. Forced out of his cabin by the rising water, Carter helped eighteen women and children into a large hackberry tree. They would spend three days and nights in that tree.

One woman kept going to sleep and falling into the water, forcing Carter to dive in and rescue her until he finally tied her to a limb. When people became thirsty, Carter would lower his shoe into the water with a small rope, then bring it up. It wasn't exactly drinking wine out of a slipper. He had to repeat the process until everyone's thirst had been slaked. Finally, the party was rescued by M.D. Fields and others in skiffs. They were brought to a camp on high ground on the east side of the Brazos across from Richmond. There, an acquaintance named August Meyer asked if he could bring Carter anything. After three days in a hackberry tree drinking floodwater from a shoe, Carter answered: "I want a half and half gin and whiskey, quick."[5]

Other people along the Brazos were also rescued from treetops and rooftops.

Two other accounts of the flood came from Mennonite P.S. Warkentin, a Fort Bend County farmer. Warkentin wrote two letters to the *Mennonitische Rundschau* concerning the flood. The first was dated June 29 from Richmond:

> It has been a while since I wrote for the "Rundschau." I do not know when this piece will be mailed because it rained here so much. The creeks are up and a person cannot get into town. It is raining again right now.
>
> Because it was dry for several weeks we gave up on the corn. Had there been rain three weeks ago things would be different today. Now the wind pushes the corn to one side and then the other. The roots come loose in the rain-soaked soil. The rain began at 11 P.M. on the 26th and continued until 1 P.M. on the 28th. That is 38 hours with hardly any let up. At times it poured buckets. Our farmland was under water with only the tops of the cotton and corn sticking out. The prairie was covered with five inches of water. Yesterday the creek close to our house spilled over its banks, something that has not happened since our arrival in Texas. The water continued to rise until this morning. It reached the animal barn within a distance of two feet. This morning it slowly began to recede. Right now (noontime) it is beginning to rain intermittently again. It is raining hard enough for the water to rise. . . ."[6]

Warkentin's second letter, published a week later, was dated July 10 from Richmond:

> I am writing quickly a few lines since I happen to be in town. Getting here had its challenges. Many of you probably already heard of the great flooding when you read this. The Brazos River rose about 40 feet and the bottom lands which are between 4 and 8 miles wide have become a lake. I was here this past

Friday when the water still rose. Today it receded 18 inches. What does one see or not see? It is impossible to estimate the damage. The best cotton and corn harvest in years was anticipated. Now everything is gone. In some places the water is 20 feet on the cotton fields. In this town 1,700 inhabitants have to be fed. How many lives were lost to the flood? No one knows yet. In Brookshire on the 7th of this month 13 lives were lost as well as numerous livestock. It is very sad. In our [Mennonite] settlement [near Fairchilds] we are "all right" so far. Our harvest suffered, to what extent we do not know yet. . . .[7]

Heavy crop damage came in McLennan, Falls, Robertson, Milam, Brazos, Burleson, Grimes, Washington, Waller, Fort Bend, and Brazoria Counties, as well as other areas. Whole trees were washed down through the river bottoms. Thousands lost their homes. Livestock drowned, farm equipment was destroyed, and many of the houses that weren't washed away were left uninhabitable. In spots east of Hearne, the flood deposited five feet of sand on top of the rich bottomland. More importantly, by the time the floodwaters had passed out into the Gulf of Mexico by the middle of July, two hundred eighty-four people had drowned along the Brazos basin, including twenty-four in Robertson County.

Property damage was estimated at between eight and ten million dollars.

South of Waco, the Brazos flooding had been the worst in forty-seven years. But Texas farmers quickly bounced back. The *Climate and Crops: Texas Section* report for July 1899 reported: "Some of the replanted cotton is coming up nicely."[8]

## Chapter 9

# *The Galveston Hurricane of 1900*
# *The Storm*

### *"Darkness is overwhelming us"*

Galveston, Wednesday—Galveston has been the scene of one of the greatest catastrophes in the world's history. The story of the great storm of Saturday, September 8, 1900, will never be told. Words are too weak to express the horror, the awfulness, of the storm itself: to even faintly picture the scene of devastation, wreck and ruin, misery, suffering and grief. Even those who were miraculously saved after terrible experiences, who were spared to learn that their families and property had been swept away, spared to witness scenes as horrible as the eye of man ever looked upon—even those can not tell the story.[1]

The Thursday, September 13 issue of the *Galveston Daily News* was the first complete edition since the hurricane struck. Smaller editions, printed on hand presses, had been turned out Sunday through Wednesday. Sunday's edition had been a list of the dead, with a note that the list would be updated as more deaths were verified.

The headline said it all.

STORY OF THE HURRICANE WHICH SWEPT GALVESTON
Loss of Life Is Estimated at Between 4000 and 5000—Not a
Single Individual Escaped Property Loss—The Total

Property Loss from Fifteen to Twenty
Million Dollars.

Even those estimates were low.

Property losses have been estimated at $30 million to $40 million in 1900 dollars. Adjusted to 1996 dollars and using a 1915 cost adjustment base because none is available before that year, the Category 4 storm caused $809,207,317 in damages. The death toll is usually estimated at six thousand to eight thousand, although it might have reached ten thousand to twelve thousand. And it could have been worse. Some twelve thousand to twenty thousand fled the island before the storm struck with full intensity.[2]

This in the fourth-largest city in Texas with a 1900 population of 37,789.

The 1900 storm remains the worst natural disaster in United States history.

It wasn't that Galveston had been immune to hurricane destruction. The October 3, 1867 storm—the first million-dollar hurricane—had frightened many residents. Meteorologist C.G. Forshey noted: "The sea appears nearly all over town. Looking west and north—cannot see far. (What a morning for funerals!) . . . The pestilence cannot pause for the tempest! (Have they combined for the utter desolation of our fair city?)"[3]

But Galveston weathered that storm, as well as the 1875 and 1886 hurricanes that transformed Indianola into a ghost town, and others. So many ignored the warnings in September 1900. *Tribune* editor Clarence Ousley said: "Many were unafraid or judged their houses sufficiently strong. There had been high waters before, notably in 1875 and 1886, when the effect was mainly discomfort and wrecked fences."[4]

The tropical storm had originated in the Cape Verde Islands on August 31. It moved across the Caribbean and gained hurricane status on September 5 as it entered the Gulf of Mexico. A hurricane watch was posted along the Gulf to New Orleans on September 6, when the storm was six hundred miles east of Galveston. The next day the watch was extended into Texas. On September 7 the *Galveston Daily News* reported that the storm was northwest of Key West, Florida. On September 8 the newspaper ran a one-paragraph article reporting storm damage in Mississippi and Louisiana. The story was out of date. By the time readers sat down with their newspapers Saturday morning, the hurricane was barreling toward their city.

Doctor Isaac Monroe Cline, not quite thirty-nine years old and an eighteen-year Weather Bureau veteran, had raised storm warning flags above the U.S. Weather Bureau station. He and his brother Joseph, a Weather Bureau meteorologist, worked late Friday night before retiring to Isaac's two-story frame house at Avenue Q and 25th. It started raining after midnight. The Clines woke up around 4 A.M. and noticed a high tide. Isaac went to the beach to observe the tide and warn people of the storm. Joseph went to the office to file a report to Washington. An hour later, Isaac reported tides four-and-a-half feet above normal. At 5 A.M. the first telegraph was sent to Washington. The Clines would continue their reports at two-hour intervals until the wires were dead.

One telegram read:

> Unusually heavy swells from the southeast. Intervals one to five minutes. Overflowing low places south portion of city three to four blocks from beach. Such high water with opposing winds never observed previously.[5]

At 11 A.M. Joseph Cline telegraphed Washington that the barometer was at 29.417 and falling, the temperature was 82.8 degrees, the wind was thirty miles per hour from the north, and rain was falling heavily.

Yet people still came to the beaches. Some had come by train from Houston to watch the waves crash onto the sand. Isaac Cline took a horse-drawn cart or buggy and went up and down the coast, warning beachcombers to go back to the mainland. Few listened. Around 1 P.M. Ohio native Lloyd Fayling, managing a newspaper syndicate in Galveston, hitched a ride in a buggy down to the beach to watch the storm. Five years later, Fayling recalled:

> ... we found water from M street south as deep as the hubs of the buggy. People were already somewhat alarmed at the situation, but no one thought that the storm was going to be anything more than one of our

usual damp spells. My acquaintance left me a block from the beach to bring some ladies into the centre of the city, and I waded and swam the rest of the way to the Midway Road to get a good view of the sea. Within a few minutes of my joining the group standing in the wrecked street, O'Keefe's bathing pavilion and half of the Pagoda went down. Murdoch's pavilion was also going to break up. The crowd at the beach knew me very well, as I had been swimming several hours a day there all summer, and they began to chaff me good-naturedly about losing that new bathing suit of mine, which was of a particularly vivid color and attracted some attention. I told them in a bantering way that it had probably gone out to sea, and then one of the Pagoda attendants came along and said that it was still in the building. I suggested that he go and get it, but as he seemed to doubt that he would ever come back, I went there and got it myself. The Pagoda is a building extending out into the sea on piling, with a long walk supported also by pilings connecting it with the beach. From the Pagoda I got a close view of the sea, closer than I could have got from the beach, and as the waves were wetting me while out there, I noticed their phosphorescent color, as well as the fact that the wind was blowing a gale from the north while the waves were running against the wind instead of with it. I had seen these same signs in the West Indies while there on military service, and had found that they meant a hurricane. I then realized that something was going to drop, and started for town, bathing suit in hand, and informed all the people I met that they had better get into the higher parts of town.[6]

The Pagoda and boardwalk soon were pounded into driftwood. A steamship was torn loose from its moorings and took out three bridges. Wind and waves increased. Houses on the beach began falling, slamming into other houses, and creating a domino effect.

By now people realized the danger and began seeking shelter. But it was too late to leave the island.

> One street of buildings would go down. That would be next to the gulf. The timbers were hurled against another street. It would go down. The debris of the two would attack the third. The three would attack the fourth, and thus on till Q street was reached. Here the mass lodged.[7]

By 3 P.M. tide water covered more than half of the city. An hour later Galveston was covered by water one to five feet deep.

At 3:30 P.M. Isaac Cline handed his brother a message to telegraph to Washington, saying the hurricane would kill many people and help would be needed in Galveston. Despite waist-deep water, Joseph reached the Western Union office to learn that the telegraph lines were down. Instead, he telephoned the Western Union office in Houston and got the message through just before the telephone went dead. Isaac Cline went home to be with his pregnant wife and three daughters. Joseph Cline arrived a short time later. About fifty neighbors had already sought shelter at Isaac's home.

At a grocery store on Avenue P, a mother and her nine children took shelter upstairs. One of her daughters went downstairs to see the water. She later told a *Galveston Daily News* reporter:

> When I went down my brother went with me and the water was half way up the counter. But that didn't scare us, because we had seen high water and heard the winds before. Well, we went back and in a few minutes we were down again. Then the counter was floating. Brother said not to tell mother, but I did. Then we saw a house tumble down and we heard people crying. We got scared then and me and mamma prayed. We prayed that one of us would not be drowned and the little children were not drowned because one of us would have to be their mother.[8]

Yet some people still weren't scared. Recalled a merchant:

> We had passed through the terrible storm of 1875, and had lived. Since then on the island has been

raised five feet or more. Why should we not have felt easy? But when the wind and waves began to show their fury, when I saw these extra five or more feet covered by a raging torrent which raced hither and thither, I felt that the end had come. Up the waters came about the fence—up they came and covered the hedge. Up they came and knocked at the door. Yet I still thought the end would be reached. We had been told that the height of the storm would be at 2 [?] o'clock. At 5 and 6 and 7 the waters continued to climb and the winds to take on new strength. At the last hour they were at the door. What must come, then, at 9? My heart fell then. I had peered out of the window and saw the dreadful enemy assault the house. Then agonized people were heard. It was dark and the spray sped in sheets. Yet it was light enough to see now and then. People in boats and wading came along. Their houses were gone. Mine rocked like a cradle, and I felt the end had come.[9]

For many, the end had indeed come.

The first death was reported at Rietter's saloon on the strand. Stanley G. Spencer, Charles Kellner, and Richard Lord had been sitting on the first floor, "making light of the danger," when the roof caved in. All three were killed, and others were trapped and injured. A waiter was sent for a doctor, but the waiter only made it to 21st Street before he drowned.

One of the most moving accounts of the storm was written at the John Sealy Hospital, probably by a nurse:

A.M.

It does not require a great stretch of imagination to imagine this structure, a shabby old boat at sea. The whole thing rocking like a reef surrounded by water, said water growing closer, even closer.

Have my hands full quieting nervous hysterical women. 12 - Noon. Things beginning to look serious. Water up to the first floor in the house, all over the basement of the hospital. Cornices, roofs, window lights blinds flying in all directions.

-----

Noon.

The scenes about here are distressing. everything washed away. poor people trying, vainly to save their bedding, & clothing. methinks the poor nurses will be trying to save their beds in short order. Now flames in the distance.

It is all a grand, fine sight, our beautiful Bay, a raging torrent.

3 P.M. Am beginning to feel a weakening desire for something "to cling to." should feel more comfortable in the enbrace [*sic*] of your arms.

You hold yourself in readiness to come to us should occasion demand.

-----

Darkness is overwhelming us, to add to the horror. Dearest - I - reach out my hand to you. my heart - my soul.[10]

The wind had averaged eighty-four miles per hour and had reached one hundred miles an hour before the anemometer blew away around 5:20 P.M. The wind also destroyed the rain gauge. Caskets floated from graves in the cemeteries. Wind estimates would reach one hundred and twenty-five miles an hour. Slate shingles were blown off buildings and turned into missiles, as were bricks and lumber. Some Galvestonians seeking shelter were decapitated by these flying objects.

Lloyd Fayling remembered:

A number of people were caught very close to me during the night with bricks and masonry. One poor fellow had his whole face hashed off with a piece of slate only a few feet from me, just as he was within a few steps from the building and safety. A dozen times bricks and slate missed me by only a few inches, and I

only had a narrow margin on several chimneys, but I
seemed to be particularly in luck and scarcely got a
scratch.[11]

At the St. Mary's Cathedral, Father James Kirwin and other
priests and staff members were forced upstairs by flooding waters.
A bell, weighing two tons, was blown from its anchor and crashed
onto the tower floor. Father Kirwin looked outside to see a terrified
horse galloping through the water only to be killed by a flying timber. Bishop Nicholas Gallagher, motioned toward the assistants
and told Kirwin: "Prepare these priests for death."[12]

One man said he felt "complete resignation" by then. "Some
were frightened and simply shrieked and laid hold of anything that
would relieve them from the embraces of the water. Some were
frightened and prayed for mercy. Some were frightened into dumb
resignation, partaking of dumb indifference."[13]

Lloyd Fayling, however, wasn't resigned about anything.

After leaving the beach earlier in the afternoon, he waded
through water averaging five feet deep and dodged fallen, live
wires. By the time he reached the YMCA building, "masonry, bricks
and slate were coming pretty thick" and the wind was blowing
women ten to twenty feet in the water. Fayling helped the women
into the YMCA building and continued on to his office at 21st and
Market, where he put on his bathing suit and "a pair of stout Turk-
ish slippers."

In the Gill and League Building, Fayling found "a hungry looking crowd," so he swam across the street to the Four Seasons
restaurant. When an employee told Fayling he couldn't be served,
Fayling replied that if he didn't get food, he would drown the
employee. Fayling got an "armload of provisions."

That's not surprising. Lloyd Fayling probably cut an imposing
figure. He had served as a deputy United States marshal during the
Chicago strikes and as an undercover agent in the Appalachians
after moonshiners. He was a spy and a captain for the *Junta* in
Cuba in 1895, was captured and escaped, and raised a company of
U.S. volunteers during the Spanish-America War in 1898.

Not only did Fayling bring food across the street, he also rescued two men, one in his seventies, the other about twenty, from

drowning. There, the refugees waited out the storm. Recalled Fayling:

> Things got worse every moment. The wind rocked our place in all directions, and the air seemed to be full of roofs, slate, masonry, telegraph poles and all sorts of things. I brought in a number of people from time to time who came by in the water, which was now running furiously through the streets, the wind having kicked up quite a sea, which was smashing the plate glass windows in the stores.

A thirty-foot-long sloop, without oars or mast and loaded with several women, children, and a man floated by, headed for the bay. The man threw a rope, trying to stop the sloop's progress, and Fayling and other men hauled the boat in, unloaded its passengers, and secured the boat to the dry building. Well, not dry. ". . . we only had two inches of rainwater which had leaked in, as our roof had blown away a few minutes previous, and there was only the flooring above to keep out the rain," Fayling wrote.

In the building, Fayling and two doctors brought in forty-three people. All survived.[14]

Others managed to escape death. One soldier said he had been blown into the Gulf and floated all night on a door. Another said he swam with a cow for three miles. The cow finally sank, but the soldier swam the rest of the way to shore.

Few were so lucky.

A storekeeper named Mutto had rescued his neighbors in a one-horse cart, hauling load after load to a fire company house. "On three occasions the cart load of human beings, some half dead, others crazed with fright, was carried for blocks by the raging currents, but he landed all of them safely, even to his last load, when he met his death. As he attempted to pass into the building on his last time the fire house succumbed to the wind and collapsed. Some of the wreckage struck Mutto and he was mortally injured."[15]

A woman took her baby from her house, but a falling beam hit the child on the head and killed the five-month-old instantly. Eighteen people took shelter in Grother's grocery store, but the building was washed away and all inside presumed dead.

At the Lucas Terrace apartments, twenty-three-year-old Daisy Thorne had carried her five cats to the second-floor parlor to escape the rising water. Twenty-two people would take shelter in the Thorne apartment. It turned out to be an excellent choice. Of the sixty-four rooms in the building, only Thorne's bedroom would remain standing.

Lighthouse keeper Harry Claiborne had fled into the Bolivar Point Lighthouse shortly after the water began rising. Before long, people were pounding on his door. Reports give the actual number of refugees in the lighthouse from fifty to one hundred and fifty. People began climbing the steps to escape the rising tide. They ran out of water and tried filling buckets outside to catch rainwater, but the wind blew too much saltwater into the buckets, rendering the water undrinkable. The tower swayed in the wind, and the door disappeared under thirty feet of water, but the lighthouse survived the storm.

The two-hundred-twenty-foot-high tower above St. Patrick's Church blew over onto the streets, but the noise of the storm drowned out the crash. The church was soon destroyed. The north wall at the Ursuline Convent and Academy fell. A two-hundred-twenty-foot-long section of trestle was blown into Isaac Cline's house. The house began to fall apart. The time was about 7:30 P.M.

People inside sang, cried, prayed. Isaac Cline tried to grab his wife and six-year-old daughter but found himself swept into the water with them instead. Joseph Cline took hold of Isaac's two other daughters and threw himself backward through a window. The three floated on the top of the wrecked home. Isaac somehow made his way to the surface and saw his youngest daughter. His pregnant wife, however, remained trapped below and died. Efforts to find her proved fruitless, so Isaac drifted along with his daughter. Thirty minutes later, he spotted his brother and two other girls. The Clines managed to float amid the debris and surging water.

They dodged a floating, wrecked house and climbed aboard it, drifting out to sea. For two hours, the Clines fought to stay alive and afloat on their makeshift raft. At one point, Isaac pulled a girl from the water. At first he thought it was his six-year-old daughter. Soon, he realized the girl was a total stranger. Around 10:30 P.M. the strange raft had drifted back toward town and the Clines spotted a two-story house still standing. The people inside the house pulled Isaac, Joseph, and the three girls inside through a second-story window. They had traveled back to their own neighborhood.

Weather Bureau meteorologist John D. Blagden, on temporary assignment in Galveston, wrote on September 10:

> Dr. Cline placed confidence in the strength of his house. Many went to his house for safety as it was the strongest-built of any in that part of the town but of the forty odd who took refuge there less than twenty are now living.[16]

Blagden stayed at the Weather Bureau station throughout the storm. "It was in a building that stood the storm better than any other in town, though it was badly damaged and rocked frightfully in some of the blasts," he wrote. Blagden also noted, "I have seen many severe storms but never one like this."

The most heartbreaking story of the storm occurred at the St. Mary's Orphanage.

Sister Elizabeth Ryan had collected food at St. Mary's Hospital that morning and returned to the orphanage, which consisted of two dormitories, three miles west on the beach, despite pleas to wait out the approaching storm at the hospital. As the storm intensified, Sister Elizabeth and the other sisters moved all of the children into the newer, stronger girls dormitory and stayed in the chapel. The sisters and ninety-three orphans sang "Queen of the Waves" to try to calm the frightened children. Around 7 or 7:30 P.M. a five-foot tidal wave surged through Galveston. Rising water sent the sisters and children upstairs, where they saw or heard the boys dorm collapse.

Still, the water rose. The high-water mark would reach 15.7 feet. The sisters had a worker bring clothesline from a storeroom. They cut the rope into sections. Each sister secured the rope around her waist, then around six or eight children. Some of the older children climbed out of a window and onto the roof. One sister held two small orphans in her arms, promising the terrified youths she wouldn't let go.

Before 8 P.M. the roof of the girls dormitory collapsed.

69

Three boys—fourteen-year-old Francis Bolenick and Albert Campbell and Will Murney, both thirteen—somehow escaped. All of the sisters and the rest of the children were crushed to death or drowned. Ten sisters had worked at the orphanage, although Father Kirwin said that one had survived by leaving in a wagon and taking shelter with a family on the island before the storm reached its full fury. Other accounts say all ten perished. Whatever the number, the bodies of two of the sisters were found across the bay. One held two small children tight in her arms. Other corpses were also found, the drowned children still fastened to the sisters by clothesline.

The three surviving children clung to the top of a tree caught in the masts of a wrecked schooner. When Campbell said he was drowning, Murney tied him to the tree with a piece of rope he had found. They drifted all night until the tree broke loose and the boys drifted to shore.

The orphanage was washed away.

Father Kirwin recalled:

> I have been out to where the asylum stood, and have tried to find traces of it. There is absolutely nothing, unless it be a few scattered bricks. The asylum was not far from the beach. It was in that part of the city which was swept clean. The structure was large and strongly built. We have been able to find scarcely any part of it. At a distance of two miles down the island the other day I came upon the contribution box, which was in the parlor of the asylum. There was still upon it the inscription, "Remember the Orphans."[17]

Around 10 P.M. the tide began to recede. By 3 A.M. Fayling noted, "the wind had gone down to merely a strong gale. It seemed like a calm after the night's cyclone."[18]

The hurricane was over. But when dawn broke, the Galveston survivors would realize that their nightmare had only just begun.

*Chapter 10*

# The Galveston Hurricane of 1900
# The Aftermath

## *"I have no more family"*

**O**n Sunday morning Lloyd Fayling checked on some friends, who had survived the hurricane, and made his way downtown. Store windows had been broken, and he spotted "suspicious looking people" going through the debris. "It was not yet daylight," he wrote, "but looting had already begun." Downtown, he found Police Chief Edwin N. Ketchum and asked if he could be of any help. Ketchum commissioned Fayling as a sergeant—on a damp envelope from the Tremont Hotel. Fayling would command a makeshift militia that began patrolling the streets. "Everything was chaos," Fayling recalled. "It was the worst looking town I ever saw. . . ."[1]

Fayling's orders included:

"Close all saloons in town. If a man opens up again and sells liquor after being closed, arrest him.

"Second: Shoot anyone caught looting the dead or desecrating corpses in any way. If anyone resists your authority, shoot. Be very careful not to interfere with good citizens in any way, but investigate all suspicious characters."[2]

It's difficult to estimate just how much looting took place in the devastated city.

"A great deal of looting has been going on," the *Galveston Daily News* reported Monday, September 10, "and others who have not indulged in this practice have gotten supplies from the relief committee without working."

71

John D. Blagden wrote:

> The city is under military rule and the streets are patrolled by armed guards.
>
> They are expected to shoot at once any one found pilfering. I understand four men have been shot today for robbing the dead.
>
> I do not know how true it is for all kinds of rumors are afloat and many of them false.[3]

Fayling himself was asked in Chicago a few weeks after the storm to comment on the work of the body robbers. He answered only:

"We maintained public order at all costs."[4]

Yet one account quoted Fayling telling his men: "Shoot them in their tracks, boys! We want no prisoners!"[5]

> In every instance the pockets of the harpies slain by the United States troops were found filled with jewelry and other valuables, and in some cases, notably that of one Negro, fingers were found in their possession which had been cut from the hands of the dead, the vampires being in such a hurry that they could not wait to tear the rings off.[6]

The *Galveston Daily News* reported September 12 that eight black men were shot for looting on Monday evening, five of them shot by one man. "They were in the act of taking jewelry from a dead woman's body, the soldier ordered them to desist and placed them under arrest. One of the number whipped out a revolver and the soldier shot him. The others made for the soldier and he laid them out with four shots." Fact or fiction? Who knows?

One city official said: "Over 100 ghouls were shot Wednesday afternoon and evening and no mercy was shown the vandals."[7]

Yet Galveston had more than looters to worry about. More than three thousand six hundred homes had been destroyed, a third of the island wiped out, and the rest battered into near oblivion. "The scenes of desolation were awful," Alice Pixley said. "For three miles, in a district which had been thickly settled, not a house was standing except one or two. Masses of timbers were piled up everywhere, and hundreds of dead bodies were to be seen."[8]

Looking northwest on Nineteenth and O after the Galveston hurricane of 1900.
*Texas State Library & Archives Commission*

A clean sweep. Little is left after a hurricane
swept through Galveston in September 1900.
*Texas State Library & Archives Commission*

Dazed by the storm, people weren't sure how many had died. Father Kirwin found a group about to cross the bay in a boat to spread the word of what had happened on the island. He told them:

"Don't exaggerate; it is better that we underestimate the loss of life than that we put the figures too high, and find it necessary to reduce them hereafter. If I was in your place I don't believe I would estimate the loss of life at more than five hundred."[9]

That turned out to be a gross underestimate, but Galveston remained in shock on Sunday.

John Hayes Quarles described the horror in the *Houston Daily Post*: "It was horrifying. People who had never before viewed dead, and those who had never seen anything but the remains of dear friends and relatives, were brought face to face with the calamity, and their nerves were not fitted for the sight which met their gaze."[10]

As the *Galveston Daily News* reported September 16: "Sunday after the great hurricane it took the people of Galveston a whole day to realize what had happened. In fact, Monday was far spent before it dawned upon many citizens how widespread and awful the catastrophe really was."

Lloyd Fayling later told a reporter:

> Two days before the flood I took dinner at the home of one of Galveston's leading citizens—magnificent home, two beautiful daughters and several little tots—the kind you can't help but love; they played around my knee, we were all happy and—well, the next time I saw them—the entire family—they were almost unrecognizable corpses in the ruins of that beautiful home. But that's nothing. There were hundreds of just such cases as that.[11]

On Sunday afternoon, Mayor Walter C. Jones, former Congressman Miles Crowley, and other citizens met at the Chamber of Commerce to form an emergency committee as well as committees to oversee finance, correspondence, hospital, burials, and general relief. Ben Levy drew the horrible task of burial committee chairman, "charged with collecting and burying the bodies of all dead humans and animals." Also on the committee were J.H. Stoner, F.P. Habine, Frank Sommers, Sterling Norman, and M.F. Wirf.

The law requiring a coroner's ruling of death was suspended to expedite the burial process. Apparently, they still thought they had only four hundred to six hundred dead. Bodies were still being brought to the morgue, many hauled in wagons that had been used Saturday for transporting groceries, food, beer, and grain.

Police Chief Ketchum said food should be seized from all wholesale groceries and flour mills—they would be paid for it—and the food distributed by the committee. Healthy men who would not work would not be fed.

"Hundreds are busy day and night clearing away the debris & recovering the dead," John D. Blagden wrote. "It is awful. Every few minutes a wagon load of corpses passes by on the street."[12]

Temporary morgues were set up, but as more bodies were discovered and temperatures rose, city officials realized they needed to take drastic measures to dispose of the dead. By 9 A.M. any attempt to identify the corpses ended. Bodies were now hauled to a barge to be taken out to sea.

Even that proved difficult.

> There is shortage of horses to haul the dead, and there is a shortage of willing hands to perform the gruesome work. . . . it would be impossible to bury the dead even in trenches, and arrangements were made to take them to sea. Barges and tugs were quickly made ready for the purpose, but it was difficult to get men to do the work. The city's firemen worked hard in bringing bodies to the wharves, but outside of them there were few who helped.[13]

Father Kirwin recalled:

> Some of our best men took the lead in this, to set the example. They went right out and helped pick up the bodies. But hard as we worked, the more there seemed to be. It soon became so that men could not handle those bodies without stimulants. I am a strong temperance man. I pledge the children to total abstinence at communion; but I went to the men who were

handling those bodies, and I gave them whisky. It had to be done.[14]

"My God!" one of the workers shouted on Monday night. "Don't bring any more!"[15]

It's understandable. *Tribune* editor Clarence Ousley noted: "Human nature has its limitations. The men in the morgues and at the wharf sickened and recoiled. Fresh recruits were brought in, some volunteers, some impressed at the point of the bayonet."[16]

Lloyd Fayling rounded up several men and forced them to help in the burning of debris or the loading of bodies. The mayor ordered Fayling to go to Father Kirwin, who explained the situation at the barges. Fayling wrote:

> . . . we marched to the foot of Tremont street, taking every able-bodied man, white or black, met with, and forced them at the bayonet point to assist in the awful work. These poor fellows were only kept up on whisky, which was given to them by the goblet full, but I did not see any drunken men among them. The stench was terrible, and the work was so disgusting that almost every moment these men were forced to stand aside a few moments, their stomachs rebelling at the terrible task. Men would say "For heavens sake don't make me do that! I won't go, you can shoot me if you want, but I will not and I can not." Our only answer was "Load with ball cartridge, take aim —" and fortunately we never had to go any further. They always threw up their hands and went to work. I do not know whether I would have shot them or not. But of course, as the orders were to do so, I think I would.[17]

Three barges carried seven hundred dead out Monday night. The barges were taken eighteen miles into the Gulf, and under guard, the impressed and volunteer workers fastened weights to the corpses and threw the bodies overboard. An estimated two thousand three hundred bodies were dumped at sea.

A few days later, many of the bodies, improperly weighted, washed back on shore. The burial committee adopted another

Dead bodies can be seen on a railroad barge on the
day after the horrible Galveston hurricane of 1900.
*Texas State Library & Archives Commission*

plan. Bodies were drenched with kerosene and burned. The
funeral pyres continued for more than a month.

Mrs. Fannie B. Ward, a special assistant with the Red Cross,
described one of the cremations in her report:

> Boards, water-soaked mattresses, rags of blankets
> and curtains, part of a piano and the framework of
> sewing-machines, piled on top, gave it the appearance
> of a festive bonfire, and only the familiar odor
> betrayed its purpose.
>
> "Have you burned any bodies here?" I inquired.
> The custodian regarded me with a stare that plainly
> said, "Do you think I am doing this for amusement?"
> and shifted his quid from cheek to cheek before
> replying.
>
> "Ma'am," said he. "This 'ere fire's been goin' on
> more'n a month. To my knowledge, upwards of sixty

bodies have been burned in it,—to say nothin' of dogs, cats, hens, and three cows."

"What is in there now?" I asked.

"Wa'al," said he meditatively, "It takes a corpse several days to burn all up. I reckon that's a couple of dozen of 'em—just bones, you know,—down near the bottom. Yesterday we put seven on top of this 'ere pile, and by now they are only what you might call baked. To-day we have been working over there (pointing to other fires a quarter of a mile distant), where we found a lot of 'em, 'leven under one house. We have put only two in here to-day. Found 'em just now, right in that puddle."

"Could you tell me who they are?" I asked.

"Lord! No," was the answer. "We don't look at 'em any more'n we have to, else we'd been dead ourselves before to-day. One of these was a colored man. They are all pretty black now; but you can tell 'em by the kinky hair. He had on nothin' but an undershirt and one shoe. The other was a woman; young, I reckon. 'Teeny rate she was tall and slim and had lots of long brown hair. She wore a blue silk skirt and there was a rope tied around her waist, as if somebody had tried to save her."

Taking a long pole he prodded an air-hole near the centre of the smoldering heap, for which now issued a frightful smell, that caused a hasty retreat to the windward side. The withdrawal of the pole was followed by a shower of charred bits of bone and singed hair. I picked up a curling, yellow lock and wondered, with tears, what mother's hand had lately caressed it.

"That's nothin'," remarked the foreman. "The other day we found part of a brass chandelier, and wound all around it was a perfect mop of long, silky hair—with a piece of skin, big as your two hands, at the end of it. Some woman got tangled up that way in the flood and jest na'cherly scalped."[18]

Joseph Johnson of Austin had arrived in Galveston on Friday. When he left Tuesday, he said, "the stench from decaying human bodies was simply terrible and almost unbearable." Johnson also noted that "It would take 5,000 men one year to clear the streets and town of Galveston, so complete is the ruin."[19]

By Tuesday, Galveston realized its early estimates of the dead had been way low. "Conservative estimates place the number of dead in the city at 2,000," the *Galveston Daily News* reported September 12.

Even that number didn't come close to the truth.

At noon Tuesday, Mayor Jones had posted a notice on the city streets:

> To the Public: September 11, 1900. The city of Galveston being under martial law, and all good citizens being now enrolled in some branch of the public service, it becomes necessary to preserve the peace that all arms in this city be placed in the hands of the military. All good citizens are forbidden to carry arms, except by written permission from the Mayor, Chief of Police or the Major commanding. All good citizens are hereby commanded to deliver all arms and ammunition to the city and take Major Fayling's receipt.
> Walter C. Jones, Mayor[20]

The city was placed under martial law, and the governor sent General Thomas Scurry, adjutant general of the State Volunteer Guard to Galveston. He arrived on Tuesday evening. Lloyd Fayling, exhausted after hours without sleep, requested to be relieved of his commission. Scurry approved the request and added an uncustomary: "Your services have been most worthy."

Jones also sent out an appeal for help:

> It is my opinion, based on personal information, that 5,000 people have lost their lives here. Approximately one-third of the residence portion of this city has been swept away. There are several thousand people who are homeless and destitute—how many there is no way of finding out. Arrangements are now being made to have the women and children sent to

Houston and other places, but the means of transportation are limited. Thousands are still to be cared for here. We appeal to you for immediate aid.[21]

Removing the dead was one thing, but Galveston also had to help the living.

"Men are just beginning to realize that the living must be cared for," the *Galveston Daily News* reported September 14. "It is now the supreme duty."

"There is not room in the buildings standing to shelter them all and hundreds pass the night on the street," John D. Blagden wrote. "One meets people in all degrees of destitution. People but partially clothed are the rule and one fully clothed is an exception."[22]

By the time the Red Cross arrived eight days after the hurricane, Fannie Ward noted, "The most we could do for the grief-stricken survivors was to mitigate in some degree their bodily distress."[23]

> The wounded everywhere are still needing the attention of physicians, and despite every effort it is feared that a number will die because of sheer impossibility to afford them the aid necessary to save their lives. Every man in Galveston who is able to walk and work is engaged in the work of relief with all the energy of which he is capable. But despite their utmost endeavors they cannot keep up with the increase of the miserable conditions which surround them. Water can be obtained by able-bodied men, but with great difficulty. Dr. Shaw of Houston, who is busily engaged in the relief work, said tonight that there were 200 people at St. Mary's infirmary without water. They have been making coffee of salt water and using that as their only beverage.[24]

Word of the devastation spread across the country by various means. Fred Napp wrote a letter the day after the storm to Maggie Napp on the back of a piece of wallpaper. The letter read: "We had a hurricane down here on Saturday and Saturday night. It demolished about half the city. I am all right, let you know more later on. Hope that you and the children are safe. . . ."[25]

R.G. Lowe, manager of the *Galveston Daily News*, reported:

> My estimates of the loss on the island of the city of Galveston and the immediate surrounding district is between 4000 and 5000 deaths. I do not make this statement in fright or excitement. The whole story will never be told, because it cannot be told.
>
> The necessities of those living are pressing. Not a single individual escaped property loss. The property on this island is half swept out of existence. What our needs are can be computed by the world at large by the statement herewith submitted, much better than I could possibly summarize them. The help must be immediate.[26]

Some survivors went to Houston. "At times a man and his wife, and sometimes with one or two children, could be seen together, but such sights were infrequent, for nearly all who went to Houston had suffered the loss of one or more of their loved ones."[27]

Others just wanted to leave the demolished city and never return. When the trains were running again, one woman took a collie with her. She had spent the night in the storm with a child, clinging to wreckage. The collie crawled up to her and kept her warm as they drifted about the sea. Told that she couldn't take the dog on the train, the woman replied: "I'll take this dog with me if I have to pay full fare." The dog went with her and the child—and she didn't have to buy a ticket for the collie.[28]

One man in Houston, however, was determined to reach Galveston.

W.L. Love, a printer at the *Houston Daily Post*, had been in Houston during the hurricane, but his wife and son were in Galveston at Fourteenth Street and Avenue N. He boarded a train, but washouts forced it to stop eight miles from the bay, so Love walked to the shore at Virginia Point. He couldn't find a boat, so he took a railroad cross-tie and, guiding it with a stick, started for the island two or three miles away. Halfway across, a man picked him up in a boat and took him to shore.

Love found his wife and son, Sidney, alive. Water had been three feet in their home, when a Captain Lott arrived with a boat to take

Mrs. Love, Sidney, and the others in the house—Mrs. Love's grand-mother and mother—away. Mrs. Love and Sidney got in the boat, but the other women refused, saying high water was not unusual. The house was carried out to sea, and the two women were killed.

Newspapers had trouble reporting the list of the dead. "In many cases it has been impossible to ascertain the full names of victims," the *Galveston Daily News* reported. Houston and Galveston newspapers reported that Peter Boss and his wife and son had been killed in the storm. On September 18 they were found very much alive.

Help quickly came to Galveston. Clara Barton and the Red Cross learned of the disaster on September 10. Three days later, she and other Red Cross workers left Washington for Galveston. She would stay for two months.

Barton found thousands of homes washed away, others piled in ruins—"a worse than worthless mass, a menace to the safety of the remaining portion of the city."

> It was one of those monstrosities of nature that defied exaggeration and fiendishly laughed at all tame attempts of words to picture the scene it had pre-pared. The churches, the great business houses, the elegant residences of the cultured and opulent, the modest little homes of laborers of the city of nearly forty thousand people; the center of foreign shipping and railroad traffic lay in splinters and debris piled twenty feet above the surface, and the crushed bodies, dead and dying, of nearly ten thousand of its citizens lay under them.

The carnage shocked many of the Red Cross workers. Fannie Ward said "The most sensational accounts of the yellowest journals fell far short of the truth—simply because its full horror was beyond the power of words to portray." Ward went on:

> Dead citizens lay by thousands amid the wreck of their homes, and raving maniacs searched the debris for their loved ones, with the organized gangs of workers. Corpses, dumped by barge-loads into the Gulf, came floating back to menace the living; and the

nights were lurid with incinerations of putrefying bodies, piled like cordwood, black and white together, irrespective of age, sex, or previous condition. At least four thousand dwellings had been swept away, with all their contents, and fully half of the population of the city was without shelter, food, clothes, or any of the necessaries of life. Of these, some were living in tents; others crowded in with friends hardly less unfortunate; many half-crazed, wandering aimlessly about the streets, and the story of their sufferings, mental and physical, is past the telling. Every house that remained was a house of mourning. Of many families every member had been swept away. Even sadder were the numerous cases where one or two were left out of recently happy households; and saddest of all was the heart-breaking suspense of those whose friends were numbered among the "missing."

We find it hard enough to lay our dead in consecrated ground, with all the care and tenderness that love can suggest, where we may water the sacred spot with our tears and place upon it the flowers they loved in life; but never to know whether their poor bodies were swallowed by the merciless Gulf, or fed to the fishes with those gruesome barge-loads, or left above ground to become an abomination in the nostrils of the living, or burned in indiscriminate heaps with horses and dogs and the mingled ashes scattered to the winds—must indeed have been well-nigh unbearable. No wonder there were lunatics in Galveston, and unnumbered cases of nervous prostration.

Fred L. Ward, a field and financial agent for the Red Cross, helped supply ward stations with clothing, bedding, and other goods from 6 A.M. until 10 P.M. daily until leaving November 1. One man, about fifty years old, came in with his seventeen-year-old daughter to be outfitted with clothes. Afterward, Ward asked for his name and address, then asked if the man needed anything for the rest of his family.

With tears rolling down his cheeks, the man replied: "I have no more family: My wife, six children, and home were all swept away by the storm. This girl is all that I have left in the world."

Fred Ward realized the warehouses lacked enough clothing for infants, so he asked for donations. Clothes came from Massachusetts. Other supplies came in from New Orleans, Michigan, across Texas, across the United States, across the world. Ward reported that $17,341.51 was contributed to the Relief Fund.

By September 23 Clara Barton noted that the relief work had become organized and "the confusion and shock of the storm had so far passed away. . . ."[29]

But would Galveston survive? A quartermaster reported to the War Department on September 12 that "I fear Galveston is destroyed beyond its ability to recover."[30]

Clarence Ousley, however, countered that: "Galveston cannot die because its existence is a logical commercial necessity. . . . Stronger than wind or wave is the tide of commerce. . . . The harbor is here, and if every man now on the island should abandon it, ships would still come, railroads would be operated, and other men would take up the work of rebuilding."[31]

The *New Orleans Picayune* agreed, saying, "Galveston will recover." The report continued:

> It so happens that Galveston, despite the dangers of its location, is a very important shipping point for both foreign and domestic commerce, and it will continue to possess the same advantages as it had before its calamitous visitation. For years to come, every blast of wind and the hollow roar of the surf on the beach will fill the people with shudderings of anxiety and evil forebodings; but, all the same, they will continue to reside there to do business.[32]

Surprisingly, Galveston quickly recovered. Mail arrived September 12, and the city's water supply was restarted. The telegraph was back up on September 13, and banks opened September 14. Long-distance telephone service resumed on September 17, the bay railroad bridge was completed on September 21, and schools opened October 22. Martial law ended at noon September 21.

Galveston also vowed to take precautions against another deadly storm. Construction began on a seawall, seventeen feet thick at the base and standing seventeen feet above the mean low tide, to stretch for six miles along the Gulf. Contracted in 1902, the wall was completed on July 29, 1904. The city also raised every building by pumping fourteen million cubic yards of sand from the gulf.

Galveston would not become another Indianola.

The News said from the start that Galveston would rise again. Those who doubted that earlier utterance have cause to doubt and are almost as much amazed at the rapidity with which this resurrection has begun as they were appalled at the catastrophe caused by wind and waves.[33]

Built in 1894, the Grand Opera House suffered extensive damage during the 1900 Galveston hurricane but survives to this day.
*Author's Photo*

## Chapter 11

# The Goliad Tornado of 1902

### *"God seemed nigh"*

**M**ention Goliad, and most Texans will remember the fight for independence from Mexico. There, on Palm Sunday, March 27, 1836, Colonel James W. Fannin Jr. and his men, who had surrendered after the battle of Coleto, were executed by order of General Antonio Lopez de Santa Anna. More than three hundred and forty men were killed; only a few managed to escape during the melee. "Remember the Alamo!" and "Remember Goliad" became battle cries for the Texans, though the latter is often forgotten today.

There's more history at Goliad than the Fannin massacre. The South Texas town was established by the founding of Presidio Santa Maria del Loreto de la Bahia (commonly called Presidio La Bahia) and Mission Nuestra Senora del Espiritu Santo de Zuniga (or simply Mission Espiritu Santo) by Spanish soldiers in 1749. Ignacio Zaragoza was born here in 1829. As Benito Juarez's minister of war, he commanded the forces that defeated the French at Puebla on May 5, 1862. Cinco de Maya is celebrated in Texas and is a national holiday in Mexico. And then there was a dark, tragic Sunday in 1902.

It looked like rain as May 18 dawned, but nothing out of the ordinary for the thousand or so residents. They cooked, looked over their fields, socialized, went to church. As the services ended, the weather worsened. Winds blew from the southeast with gusts of forty miles per hour, and maybe even more. The sky cleared momentarily around noon, but the sky soon darkened. Buildings trembled. A few drops of rain fell.[1]

It had been a wet—and deadly—spring in much of Texas. Tornadoes struck Dallas, Ellis, Hill, Hunt, Falls, and Wise Counties on March 11, injuring sixteen, and an April 28 tornado killed eight and injured fifty-seven in Somervell County. But none could compare to the destruction about to occur at Goliad.

At 2:30 P.M. "a heavy black mass of cloud" was spotted in the northeast. Lightning flashed, and heavy thunder rolled as the cloud rapidly moved toward the city. A farmer named Johnson watched black cones dip from the clouds to the earth and pull up. Goliad wasn't tornado country, but the farmer had been to Kansas and recognized the cones as "twisters." By 3 P.M. the black cloud had reached the edge of Goliad. Heavy raindrops and many hailstones fell, along with "peal after peal of sharp, very heavy thunder with flashes of the most vivid lightning that ran over everything, giving the peculiar bluish-green white light of electricity."[2]

And then . . .

"Suddenly there broke on the ear a sound, if once heard, will never be forgotten," the *Goliad Guard* reported in its May 24, 1902, edition. "It sounded like a million-ton engine dragging a train of giant cars that had gotten away from the driver on a down grade and was rushing at a speed of thousands of miles per minute from this to some other world. The sound seemed too grand, too awful for this world."[3]

A Goliad man told a correspondent for the *San Antonio Semi-Weekly Herald* that the tornado came so suddenly that he had time only to call to his wife "when it was upon them with all its fury and terrible death and destruction."

Eyewitness J.W. Browne had been talking to friends on the gallery of the Fannin Hotel when the storm hit. He told the *Victoria Weekly Advocate*:

> Suddenly a noise came like a heavy train running in the distance. It rapidly increased in power and sound until it sounded like a million-ton engine running away. Everything turned to my eyes a dark brown or red color. Limbs of trees, debris and everything filled the air. God seemed nigh. A horrible roar, a sigh as [though] the earth were dead and the rapid dum, dum, dum, faster than you can think was over.[4]

The tornado traveled about a mile along two blocks west of the courthouse. The *Goliad Guard* reported: "While the storm was at its height, the sky, high up, was whirling round and round and looked as though filled with an immense hive of awful monsters that circle like bees. A scintillating, dull metal brown hue pervaded the zenith, the glare being unearthly in its threatening light. No sun, no star, or moon ever cast such radiance; it looked like an inferno." The hue might have been metallic dust caused by the static electrical conditions and powdered, oxidized soil.[5]

On a farm northwest of town, a fifteen-year-old girl watched the tornado with her family. Ninety years later, Gertrude Todd told the *Victoria Advocate*: "We could see something in the air, just tumbling. We said it was a horse. We knew about where it fell, so we went out there and looked. . . . It was a lady." She was dead.[6]

Eight-year-old Kate Chilton's ten-year-old frame home was destroyed in the storm. Her father, Dr. Louis W. Chilton, had gone to treat a patient in Sarco, south of town, leaving Kate with her four-year-old brother, Warren, and their mother. At 2 P.M. Kate's mother decided they should leave the house and take shelter in the nearby three-story girls dormitory, constructed of the same stone as the Presidio La Bahia, at Brooks College. Outside, Kate noticed books flying through the air as the Chiltons and three other women struggled to reach the dorm. "We got inside the building," Kate recalled in 1977, "and we held each other's hands for protection and comfort. Just as we opened a door to a closet at the bottom of an enclosed staircase, the worst and most terrible part of the storm hit."[7]

Pots, pains, chairs, dressers, and beds spun around the room in circles. "I was fascinated, but I knew that it was terrible," Kate said.[8]

The roof "was torn from the building, and everybody thrown into a heap" as the wall collapsed. Young Kate "was sucked up in the air by the wind, how high I don't know, but high enough to have a quick look at the destruction going on."[9]

Kate landed in the yard, "rolling in water," with only a scalp wound from a piece of steel. In the rubble, where the wall had collapsed onto the "heap," Warren received only slight bruises, and

Mrs. Chilton broke her pelvis and had two other injuries. Of the three other women with the Chiltons, two were killed and one broke her back but recovered.

Rob Linburg reached the Chiltons first. "I don't know how they got my mother out," Kate said. "Later, when I'd go to the meat market, he'd say, 'How's that scar on your head today?'"[10]

In less than five minutes, the tornado was over. But during that time, the winds reached more than three hundred miles per hour, perhaps as high as three hundred and sixty inside the cyclone.[11] The destruction seemed unfathomable.

According to a Weather Bureau report, "The storm was a typical tornado and presented all of the characteristic features of this destructive phenomenon."[12]

Splinters and rubble were all that remained of houses, including the Chiltons's, and the bridge across the San Antonio River, where the twister apparently touched down first, was demolished. The tornado picked up a giant steel beam from the bridge and speared it into the ground near the courthouse. At the slaughterhouse, a steam engine, weighing nearly a ton, had been picked up and set twenty feet from the slaughterhouse. The Episcopal church was severely damaged. The Methodist church, Baptist church and parsonage, and the black Methodist church were also destroyed. The latter, on Fannin Street near the San Antonio River bridge, had been filled with between forty and sixty worshipers. All were killed.

Gertrude Todd's pastor had also been killed. The tornado had skewered him with a plank that had to be sawed off at the ends before he could be buried. A child had been hurled into a well and killed. "Horses and cows, and everything that had any life at one time, were dead," Todd said. "Those who were able screamed and cried . . . they were just crying because it was so bad."[13]

The May 24 edition of the *Victoria Weekly Advocate* commented: "Frame houses were converted into avalanches of splinters, and fearful was the fate of the mortal found in the path of the death dealing storm. . . . Brick structures were carried away as easily as frame shacks."

As soon as the tornado had passed, J.W. Browne ran to the west side of town. He told the *Weekly Advocate*: "One block west of the square—the great live oaks were up rooted; two blocks—horror!

Looking at the San Antonio River bridge at the end of
San Patricio Street after the 1902 tornado in Goliad.
*Markethouse Museum, Goliad, Texas*

Many deaths and injuries occurred when the tornado
destroyed the Fannin Street Methodist Church in Goliad.
*Markethouse Museum, Goliad, Texas*

Shrieks of the wounded met the ear, the streets were a litter of dead, everything—people, cows, dogs, cats, chickens—in fact, every imaginable thing that one could think of. In company with many others, I helped all I could. The dead were on every side, white and black locked in a last death clasp to what they had seized upon."

Laura Linburg Baker recalled "seeing the trees up in the air—the houses blowing down. . . . And there were babies—little babies—lying in those prickly fences and all like that. The wind had blown them away." A correspondent for the *San Antonio Semi-Weekly Express* discovered the body of a little girl, about four years old, under a wrecked building "and many more will doubtless be found in the same district."[14]

The *Goliad Guard* was even more graphic:

> A space three hundred and fifty yards wide and a mile long, the western slope of the city, that a few moments before had been covered with pretty homes, handsome flower gardens and orchards; its streets shaded by beautiful trees, many of them giant live oaks that had withstood the storms of centuries—was now a wide waste from whose gruesome ruins came the shrieks of the wounded and the dying.
>
> Men rushed together in pairs and small parties, excitedly calling to each other; ran rapidly from one ruin to another, lifting the wounded from under the timbers, laying them down and rushing on to the next cry. 'Twas all they could do just then.
>
> The dead, and many yet quivering in the final agony, everywhere. Dying mothers shrieked for their dead and dying babes, torn from their arms by the ruthless storm and gone—they knew not whither; men bloody and staggering, were found trying to lift from their loved ones the ruins of their homes.
>
> Not a house was standing—swept as with a broom in the hands of some crazed monster demon. Only splinters and fragments were strewn around. In the eastern edge of the storm-swept region, the stone residence of R.T. Davis, proprietor of the Guard, stood like a battered and torn old fort. Roofless with its wide

galleries gone, the giant oaks around it piled in confusion over it, it was made the gathering place for
carrying the dead and dying.

When scores are dead and wounded, it is impossible to individualize. As rapidly as willing hands could
work, the wounded were carried in hacks, wagons
and every available vehicle to private houses, hotels
and public buildings. The calamity was so awful that
the people could not realize it. Whites, Mexicans, and
Negroes had alike paid tribute with their lives on the
altar of the Storm God. Many who were picked up
alive, horribly mangled, died in the arms of their rescuers, or soon after reaching a place of shelter.

People who escaped ran crying through the storm
district seeking lost ones. Dying animals set up their
mournful cries with those of their owners. Men and
women worked like heroes wiping the mud and grime
from the eyes of the dying, trying to stop the wild
wails of some suddenly orphaned little one.[15]

Some houses escaped destruction. Laura Linburg Baker said
the tornado "didn't hurt our house. We were living in the new
house then. And I'm telling you, it blew so many bedbugs and
insects into our house that we had a time getting rid of them." And
the *Victoria Weekly Advocate* noted, "as usual in great catastrophes," that the tornado "passed within a few feet of a residence,
carrying giant live oak trees, yet leaving the building untouched."

The *San Antonio Semi-Weekly Express* reported:

Those who were not in the track of destruction
were struck with awe, and it took quite a while to realize that such havoc could be wrought in such a short
space of time. Slowly they awakened to the suffering
of their relatives, friends and fellow citizens. Men's
noblest traits were in full evidence. Relief to the suffering and tender care of the dead were their first act
and thought. The message of sorrow was sent to
neighboring cities, who are now viewing with each
other in acts of kindness and charity.

The *San Antonio Semi-Weekly Express* estimated the damage at $100,000. That figure later rose to $125,000.

"Twenty-four hours ago, its inhabitants were happy and prosperous and the merry laughter of children en route to and from their usual places of gathering on the Sabbath could be heard," the *Houston Chronicle* reported May 20. ". . . It was the fearful fury of the elements to work this transformation and the dread wind to lay waste of an entire section of the city and necessitate the addition of probably a hundred or more mounds to the little cemetery in which sleeps the dead for a century back."[16]

Calls for help quickly went across the state.

Seventeen-year-old Sam Johnson saw the storm come in while he and an uncle were inspecting cattle in the morning. That afternoon, "we heard two great claps of thunder and saw great flashes of lightning. My father said he believed Goliad was being torn up, as the blackest cloud we had ever seen was moving fast." Johnson continued:

> We heard a roar like a train for about two or three minutes, which we found out later was when the worst of the storm was actually passing. We didn't think any more about it until about 5:30 when people in hacks, buggies and horseback came through rushing on their way from Runge to Goliad.
>
> My father quickly loaned out every horse he had.[17]

By 7 P.M. trains began arriving from Victoria, Cuero, and Beeville with doctors, nurses, and medical supplies. Victoria also sent the O'Connor Guards, under the command of a Lieutenant Klein, to help keep order, along with a detachment from the city's fire department. Several women from Victoria volunteered as nurses. Victoria's mayor arrived on the first train and telephoned Yorktown to send more doctors. Goliad merchants provided anything needed from their stores. Sightseers also showed up, much to the wrath of the *Goliad Guard*. "So far there have been but one drawback: different stockmen have contributed beeves, merchants and others, coffee, bread and pickles. These were given for the helpers, but a crowd that has flocked in on excursions demand food, have no money to pay for it, and are in the way. No sightseers of any sort are wanted."[18]

One report said the San Antonio River had been "sucked up into the clouds, so that the river was dry." Another said the tornado "seemed to pick up all the mud, water and slime in the San Antonio River and dump it on Goliad." Others say Dr. Chilton swam his horse across the river because the bridge had been destroyed. Either way, Chilton found his family and soon went to work with the other doctors to treat the injured.

There were several problems. Goliad had no hospital or central water supply, and many cisterns had been destroyed. The eight-year-old county courthouse became a hospital. Gertrude Todd said the dead were placed in a tin building. "And all those who weren't quite dead they put into the United Brotherhood Fellowship Hall. Of course, they were wounded and some of them died while they were in there."[19]

Dr. J.H. Reuss of Cuero was put in charge of the hospital. Committees were soon formed. N.H. Browne, Judge J.C. Burns, L.S. Seeligson, Mrs. C.D. Redding, and Mrs. L.D. Moore were appointed to a finance committee, to receive and distribute all contributions—money, clothing, and supplies. R.C. Haydon, H.H. Ewell, R.R. Stout, W.R. Taylor, and John Von Dohlen comprised the committee in charge of feeding and providing for the homeless.

The day after the storm, the finance committee issued this statement:

> To the Public: On the 18th day of May the town of Goliad was visited with a cyclone which in its effects was more horrible than the massacre of Fannin and his command. The prisoners had a certain and speedy death. Not so with all of the storm-swept district which is a bare waste. Ninety-four persons were killed. Victims of the cyclone, now have been buried, after the most acute suffering by a great portion of them. More than seventy-five remain, a great number of whom are sure to die after a painful and lingering illness.

The neighboring counties have generously contributed physicians, nurses and money. Most all of the survivors are absolutely homeless and penniless.

The citizens of Goliad are accepting visitation [of the tornado] with brave hearts and giving cheerfully of their portion. Should it be the desire of the State at large to contribute towards a mitigation of our lot subscriptions will be received and acknowledged by the committee.[20]

Victoria residents quickly contributed $1,250.

In Austin on May 19, Governor Joseph D. Sayers appealed to all Texas mayors in towns with populations of more than three thousand to send food, clothing, and money to Goliad. "To the Mayor," his plea read. "Please collect and send as rapidly as possible to county judge of Goliad supplies, food and clothing for relief of cyclone sufferers." Sayers also asked all railroads to provide free transportation for those shipments.[21]

Amputations took place at the courthouse-turned-hospital. "Here was a little girl whose legs had been amputated; there was a fond mother with hip, arm, and legs broken; here was a man with a fractured skull. . . ." the *San Antonio Semi-Weekly Express* noted. Yet "the people are orderly to an extreme." Splinters driven into bodies caused infection. Many cases of tetanus broke out. People died of lockjaw and blood poisoning. Many didn't die until early June. The correspondent for the *Express* reported the day after the tornado: "There are forty seriously injured, of whom possibly seventy-five percent will recover." And: "Many of the injured will undoubtedly die tonight."

Officially, the death toll reached one hundred and fourteen—about one out of every ten residents—and some estimates are higher. Approximately two hundred and thirty were injured. Even at one hundred and fourteen, only nine tornadoes in United States history killed more people. In Texas, the Goliad death toll is matched only by the tornado that struck Waco on May 11, 1953.

Goliad wasn't the only town that suffered.

"Many other places besides Goliad were visited by severe wind storms on the 18th, considerable damage having been reported

The tornado that swept through Goliad in 1902 left one hundred and fourteen people dead.
*Markethouse Museum, Goliad, Texas*

from Austin, Dallas, Texarkana, and San Antonio," the Weather Bureau noted. Rain drenched much of South Texas, while the roofs were blown off the Opera House and South Side college building as far northeast as Tyler. Tornadoes reportedly struck Fairfield, Thorndale, Buda, Manchaca, and Hondo; lightning killed two cows on William Adams's ranch near Alice; and a sandstorm in Bexar damaged corn and cotton, demolished a barn, and killed poultry. Heavy rains at Hearne were said to have been the most since the 1899 flood. At Luling, "a heavy dark cloud turning to a greenish hue just before reaching the city came up rapidly," resulting in a cyclone and heavy rain that damaged and destroyed trees and carried W.H. Millican's tub from his house and set it down in his front yard.

There had been strange occurrences at Goliad, too. No pieces of steel from stoves or other household items were recovered. A legend also persisted that one rooster survived the storm—with only one tail feather remaining.

More help came. By Saturday, medical help had been requested from San Antonio, Houston, Beeville, and elsewhere to relieve the doctors and nurses who had been working since the storm hit, and a Captain Barron of Victoria's O'Connor Guards left with twelve men that afternoon to spell Klein and his men. Not that there had been any trouble after the storm. "There was no looting whatever attempted; in fact, there was nothing left to loot. . . ." the *San Antonio Semi-Weekly Express* said.

The dead were buried. Goliad rebuilt. As historian Beth White writes:

> Goliad recovered, as it always had, and snapped back to normal but many people carried the scars of their injuries for the rest of their lives. For many of the following years, townspeople were very apprehensive during the month of May and May 18th was a day remembered and dreaded.[22]

## Chapter 12

# The Corpus Christi Hurricane of 1919

## *"I want to forget it all as soon as possible"*

Ten-year-old Theodore Fuller and his fourteen-year-old brother, Edwin, planned to go fishing in the morning. It was Saturday night, September 13, 1919, and reports of floundering were excellent, perhaps because the fish knew the weather was turning stormy and they needed to eat. Their fishing trip, however, was canceled before they ever left the house. "Might as well go back to sleep, Ted, there'll be no fishing today," Edwin said when Sunday morning dawned dark and wet.[1]

Fuller's father, Joseph, went to town for news on the storm. Reports had circulated during the previous week that a storm had hit the Caribbean islands. Corpus Christi, with a population of ten thousand, had weathered hurricanes before. One of the strongest had made landfall on August 19, 1916, with winds at one hundred miles an hour. That storm had caused severe damage along the bay front, with overall damages estimated at $1.6 million. The storm prompted F.L. Clendening to write:

"... the people of Rockport, Aransas Pass and Corpus Christi have experienced one of the worst storms in the history of the area and all three are nearly a total wreck."[2]

The 1916 storm had killed twenty people. Residents had been warned of the storm but for some reason didn't believe it would hit Corpus. And even if it did? Clendening wrote: "We had been made to believe that this area was storm proof."

Three years later, there had been little warning in Corpus Christi. Saturday had been a beautiful day on the beach. Storm warning flags that waved earlier in the week had been taken down.

Winds were calm. Skies were clear. The rain and high winds didn't begin until later Saturday night. On Sunday morning, however, residents realized they were in for a storm. Aransas Harbor Terminal Railway worker W.A. Kieberger told his wife at the Nueces Hotel dining room: "Now let's eat a good breakfast—we don't know when we'll get another one.[3]

People bombarded the Nueces Hotel, seeking information about trains and service calls. By 9 A.M. the phone lines were dead. An hour later, hotel officials announced that no trains would be leaving. Before long, the elevator stopped working and the lights went out. At North Beach, the U.S. Army sent trucks early in the storm to evacuate anyone wanting to come in. Most refused.

Meanwhile, Ted Fuller's father learned that the storm seemed to be south of Corpus, wasn't as strong as the 1916 storm, and the barometer appeared steady. He returned home, unconvinced, and told his sons, eighteen-year-old daughter Esther, wife Eula, and her sister "Doshie" that they should leave North Beach and go to town. Waves broke on the beach, and the tide rose, over Rincon Street and the bay, and began to cover the Fuller grounds. By the time they left, ten-year-old Ted had to wade through waist-deep water.

The Fullers took refuge in a house owned by the Atkinsons on Mesquite Street, but Joseph Fuller thought it wasn't safe enough so he decided to check out the Mayfield house about one hundred and fifty yards away on Pearl Street. Edwin Fuller went with him. They took Eula first, but as they started back, the waves and wind overwhelmed them. Joseph and Edwin gripped the side of a floating house and wound up five miles west of Corpus Christi. Meanwhile, several other residents entered the Atkinson house for shelter. As the storm intensified, twelve people were in the house. The house next door was washed off its block and floated against the Atkinson house, while the Lerick house northeast was also pushed against the Atkinsons' place. The Lericks, including nine-year-old Billy, took shelter with the Fullers and others, putting chairs on top of chests to stay above the water.

A wall separated and closed, severing three of Theodosia "Doshie" Stone Cash's fingers, but still she managed to comfort a crying boy at her side. A soldier sang and whistled to keep the refugees' spirits up. The water rose. The soldier dived through a window, climbed onto the roof, and ripped off shingles until he could help the trapped people inside onto the roof. The retreat was far from orderly. Men and women forced their way through the opening as the house slowly collapsed. Esther Fuller screamed, "Let Ted out! He's just a little boy! Look! He's a little boy! Let him out!" The soldier helped Ted onto the roof, and Esther jumped through the window as the walls merged.

Pieces of roofs became rafts. Esther climbed on board with her mother, brother, and aunt. Others in the house hung on to debris or climbed aboard other chunks of roofs. The Fuller raft overturned, and Esther, Ted, and Doshie grabbed a power pole for a while, then found another flimsy roof that waves, ten feet high, demolished. Esther and Ted were soon separated from Doshie, and screaming above the noise of wind and rain, she drowned. The Fullers found another raft, which Billy Lerick soon found. Drifting debris cut the Fullers and Lerick during the storm. By now, the only light came from flashes of lightning. Another wave washed over Esther, Ted, and Billy. Billy soon slipped into the water and disappeared.

"Pray, Esther," Ted said.

His sister answered: "I've been praying all night."

She wasn't alone.

Some twenty-five went to John McCampbell's two-story house because it sat on a terrace three or four feet above the street. They nailed a round table against the door, but the wind ripped the door from its hinges anyway. As the water rose, the refugees moved upstairs. People sang "Nearer My God to Thee." "We knew the house was going to go down," Mrs. Robert Horton recalled years later. But it didn't.[4]

J.B. Oatman's family had barricaded "a quaint and quiet home" with "its delightful view of the bay." Oatman's oldest son made his way to the house through water and driftwood and found the doors bolted. Stuck outside, he was scaling the wall to try to enter upstairs when his family spotted him and opened the door. He convinced them to leave. They had to leave the youngest daughter's gray pony, Gipsy, behind. Dressed in bathing suits, they "emerged into water five feet deep and plunged into the seething, surging waters." They found a makeshift raft and floated to the courthouse to wait out the storm.[5]

Others weren't as fortunate. Water rose about five feet in forty minutes. Before long, Water Street and the business district were flooded. The power plant was knocked out, leaving the city in darkness. The barometer would fall to 28.67 (it dropped to only 29.53 during the Galveston storm of 1900). By 4 or 5 P.M. water "got deeper than a man's height in Chaparral Street." The Bob Hall home overturned. Mrs. Percy Reid handed her five-year-old son to her husband, who swam in water twelve feet deep, carrying his son piggyback. The wind blew Mrs. Reid into the water. She couldn't swim, so she grabbed a raft—a back porch, actually—and found herself floating into Nueces Bay with many others.[6] Forty years later, Mrs. Reid recalled:

> I looked about to see how many of our crowd I could locate. I saw the old couple, Mr. and Mrs. John Hall, but they lasted only about 30 minutes before both drowned. My raft was near them when they went down. The Brooks family (Mr. and Mrs. Terrell Brooks and two children, who had been at the Hall home) also drowned. Maggie Hall (sister of Bob Hall) was struck by something and was the first in our group to go down.
>
> It was about 5 o'clock Sunday evening when we were thrown into the water. Quite a bit later I saw Mrs. Bob Hall swimming. She finally reached my raft and we were together for some time.
>
> As the wind would blow us off the raft our clothes would catch on wreckage or be sucked down by water and we would nearly drown. Then, when we struggled back onto the raft the wind would blow them worse. So we took off all our clothes and then had less trouble staying on the raft.[7]

They had to free the raft from timbers constantly. Mrs. Reid's collie, Scotch, had swam to the raft, too, and proved his worth. When strong winds overturned the raft, pinning Mrs. Reid under water, Scotch would swim under, grab her hair, and pull her back on top of the raft. When the raft capsized again, Mrs. Reid struggled to free herself but couldn't find her dog. She told the *Corpus Christi Caller-Times*:

> When he didn't appear I swung my body in the water and turned the raft over again but he wasn't there. I will never forget what a faithful friend he was and how I missed him after he was lost. That was when my suffering really began, for the warmth of the dog's body had helped me keep warm. I never came so near freezing as on that night. My teeth chattered so much that the enamel on every one of them was chipped and cracked.[8]

Around 2 or 3 A.M. after nine or ten hours in the water, Mrs. Reid landed on shore and found shelter in a cliff-side cave.

Others went to the courthouse or post office to wait out the storm. The bay eventually reached the post office, and the business district was four to six feet under water. A four-room house floated down Belden Street, and after the storm, debris hung from trolley lines. At the courthouse, where almost twenty-five hundred people took shelter, water rose in the first floor. Outside, men formed a human chain to prevent them from being swept away. Others found boats.

Inside the courthouse that night, women gave birth on the third floor, yet no child cried and no one screamed. People spent the long night speaking in hushed tones.

At the Nueces Hotel, despite being warned to stay away from windows, Lucy Caldwell, a schoolteacher on vacation, peered outside.

> ... we saw that the wires were all down, telephones were all gone, not a bathhouse was in sight, not a fishing pier, the garage near the hotel was gone with 60 cars in the bay, the concrete service station was gone, also the dancing pavilion and the bowling alley. All of the timber etc. of which these buildings were constructed was piled in front of the Nueces Hotel.[9]

Water rose in the hotel lobby, while windows and doors were smashed. At 6 P.M. the storm intensified, and two hours later, Caldwell wrote, "a mass of human beings," estimated from seven hundred to one thousand, had been "imprisoned in the hotel."

Others had gone to the Breakers Hotel, which had been converted into the Soldiers Hospital for convalescing World War I soldiers. Water flooded the first floor, taking out some of the walls and soon rose to the second floor. On Monday, the refugees were removed from third-story windows into boats.

Hit harder was the two-story frame Spohn Hospital on the North Beach lowlands.

Sister M. Paula, who arrived at the hospital shortly before the storm struck, had been awakened early Sunday morning by wind. Water was already leaking in the building. By 8 or 9 A.M. waves lapped at the doorsteps. The Weather Bureau said they were in no danger. When Sister Paula tried calling again at 11:30, the phone lines were dead.

Thirty minutes later, North Beach buildings were blowing over and the sisters were moving first-floor patients upstairs. A paralyzed woman weighing three hundred pounds was put on folded blankets and quilts and dragged up the stairs. Later the nursing staff went to chapel to be administered the Blessed Sacrament and last absolution. A short time later, Sister Paula followed Sister Thias to the northside annex to rescue a paralyzed patient. Sister Paula told the *Corpus Christi Caller-Times*: "She stepped through the door and just as she did the hall and annex blew away. Sister Thias went down before my eyes and I never saw her again."[10]

Winds, waves, and rain destroyed much of the hospital, forcing the nursing staff to keep moving forty patients. Finally, everyone crowded into a second-floor room, where one woman had to be placed in the bathtub. One survivor recalled: "Wing after wing swept away, but we kept clinging to the old part built by Dr. Spohn. We were all huddled in the rear portion of the building right over the boiler room and the Sisters were saying their prayers in concert, and I did not know how to say them their way, and I was sorry for it was a time for prayer." The Reverend Claude Jaillet, about seventy-six years old, heard confessions, gave absolutions, and prayed throughout the night.[11]

Waves washed over their feet, and the brick wall collapsed, covering the woman in the tub and sending Sister Aloysius over the side. Sister Aloysius was saved because the bricks fell on the skirt of her habit and pinned her as she dangled outside. Men pulled her back inside and cut her free from the bricks.

When morning dawned, the nuns used a red blanket for an SOS signal. Men evacuated patients and nuns in rowboats and took them to the courthouse. Surprisingly, only four people had died at the hospital, including Sister Thias. Her body was found, with black habit, veil, and rosary, shortly after the water receded. "It was a miracle that saved us," Sister Paula said.[12]

Ten-year-old Ted Fuller woke up the next morning and saw his sister beside him, their raft floating among the tops of mesquite. Ted tried to go into the water to see how deep it was, but his sister grabbed his arm. "I've held onto you too long, boy, to let you slip away from me now," she said.

Eventually, the Fullers found dry land. They had traveled eighteen miles during the storm. They eventually were found and taken to the Turner Ranch with other refugees. Later, Esther and Ted were reunited with their father and brother. Their mother, however, had drowned.

The Oatman family had something of a reunion, too. J.B. Oatman wrote: "We found Gipsy (the pony) next morning in the second story of the Williams Hotel. She is now browsing on the lawn of the kind, generous friends who are sheltering us, dispensing their hospitality so gently as to warm our hearts and enable us to look with confidence into the future."[13]

Another man awoke in a mesquite at White Point and found himself sharing his perch with several rattlesnakes. He dropped from the tree and the snakes recoiled around the limbs.

The beaches were covered with dead bodies and stunned survivors, as well as wreckage. Downtown, people used cotton bales as rafts. A cow was found stuck in a tree—and was milked from that perch.

At the Nueces Hotel, more refugees came in Monday. Lucy Caldwell wrote:

> All day the refugees poured in and many who had
> spent the night in the hotel, waded up to the bluff to

find shelter with friends and relatives. All day into the hotel poured men and women, trying to locate some relative who could not be found and who was seen last floating in the bay. One man was brought in who had floated with his wife and baby for eight hours and finally lost them. A boy came in who had seen his mother, father and five brothers and sisters drown. If I were to tell you all of them, I should use an entire tablet of paper.[14]

Waves had reached sixteen feet higher than normal, and damage estimates would reach $20,272,000. The causeway, rebuilt after washing away during the 1916 hurricane, had been destroyed and wouldn't reopen until 1921. The Aransas Harbor Terminal Railway wouldn't resume service until 1922.

Worst hit had been North Beach. All two hundred twenty-nine houses on the beach were gone, and only three buildings remained standing. Of those, only the Beach Hotel would be repaired. G.W. Greathouse, who had weathered the storm with his family in the Beach Hotel, went to his house several days later. All he found was a cut glass bowl with three pennies inside.

On La Rosita Beach, fifteen-year-old Chris Rachal and his younger brother went racing along the coast Monday morning looking for treasure washed up during the storm. Instead, they saw a body floating face down. They ran back to the house, but before they got there they saw five staggering survivors, covered with oil so they looked black. By the end of the day, seventy-five survivors were crammed inside the Rachals' home. Dead bodies were brought there, too. By the end of three days, one hundred and eight had been buried on the ranch.

Rachal recalled:

We found one [dead] woman who apparently rode the crest across the bay. She landed in a mesquite thicket on our shore. When we spotted her she was hanging from a mesquite tree by her hair. A mesquite thorn was all the way through her neck. Her feet were a good yard above the ground.

We found one woman dead on a roof. Her hair was tangled in nails so we cut it off to get her off the roof and buried. Her husband came to see us several weeks later and asked about her hair. I told him I knew where it was and went down to get it. My sister washed it in gasoline and kerosene and combed it out. The oil preserved it all those weeks.

Another man came over as we were burying people. He was looking for his mother-in-law. He found her, he said, and identified her by a missing joint on a little finger. Then he went home and found his mother-in-law alive. Several days later another man came over and identified that woman as his mother.[15]

Rachal also found a dead cow with a bridle and rope. On the other end of the rope they found a drowned woman.

D.N. Wright's father went to the courthouse to identify bodies.

More than a hundred were laid out on boxes. Thick oil made them all black. Women's long hair was matted with tar. It was hard to tell who anybody was.

The smell was terrible. I'll never forget the smell. My father was never the same again.[16]

The oil, which came from destroyed tanks, also served as a disinfectant and probably prevented sickness after the storm. Relief workers came from San Antonio. Newspapers were printed on a hand press, and later on the *Kingsville Record*'s press. The Red Cross set up operations. "Practically every pound of food in the city was in the hands of this organization," Lucy Caldwell wrote. "Nothing could be bought—money was worth nothing." The city hall, high school, and Incarnate Word Convent became feeding stations. The city hall doubled as an infirmary. Martial law was declared immediately after the storm by Mayor Gordon Boone, with Colonel Glover Johns commanding.

M.A. "Red" Harper served with the National Guard unit that patrolled the city after the hurricane. In 1978 he told the *Corpus Christi Caller-Times*: "There weren't any orders to load our rifles, but pretty soon they were all loaded. You couldn't tell who was looting and who was trying to dig out bodies."[17]

One looter, caught cutting off a dead woman's finger to get to her ring, was shot in the thigh and turned over to the provost marshal. Several other looters were shot and/or arrested. More bodies were discovered, some loaded on sleds pulled by mules. Animals—horses, cows, rats, rabbits, snakes—were dead. The National Guardsmen shot dying seagulls for target practice, until ordered by the lieutenant to stop. Harper found it hard to eat in these conditions: "I'd be hungry, then somebody would say something and I couldn't eat."[18]

Harper said:

> Nobody could have pictured a city that torn up. Houses and piers washed across the bay into houses along Water Street. They were caved in or gone. Finally lanes were cleared at intersections so people could walk across Mesquite and Chaparral streets.
>
> Every time I hear of a hurricane, I think of Corpus Christi nearly sixty years ago. Worst thing I ever went through.[19]

Lucy Caldwell would agree. She wrote her mother:

> Had one told me that I should have witnessed such a scene, I should have said "No, such a thing is impossible." . . . I want to forget it all as soon as possible. At Beeville yesterday a live wire detached by the storm killed a mule and a Mexican near the depot, but fortunately I was on the opposite side of the coach and missed it.[20]

Lucy Caldwell wrote, "The death list will doubtless reach 1,000." In reality, the death toll was placed at two hundred eighty-four, but that has been disputed. M.A. Harper said, "But from what we saw, we figured at least 600 people were killed. I believe there were nearly 300 at one time at the Court House and out on the lawn. There were others on North Beach."[21]

Wreckage shows just how hard Corpus Christi was hit by a hurricane in 1919.
*King Ranch, Inc., Kingsville, Texas*

Corpus Christi residents begin cleaning up debris and rebuilding their town after the 1919 hurricane.
*King Ranch, Inc., Kingsville, Texas*

Many people have suggested that Corpus Christi officials, seeking a deep-water port, underestimated the casualties. The Category 4 storm, which had also caused damage elsewhere along the Texas coast as well as in the Gulf of Mexico and on the Florida Keys, may have killed as many as nine hundred, so Lucy Caldwell might not have been far off the mark. But many of those deaths came at sea, where more than five hundred were lost. Still, the official number cited by Corpus Christi seems far too low. A better estimate is probably between three hundred fifty and four hundred dead.

As the 1900 storm had done in Galveston, the 1919 hurricane prompted Corpus Christi to build a seawall. The project, however, wasn't started until twenty years later. Completed in 1941, it cost about $2.2 million. A $5 million storm sewer system was installed in downtown in 1948. Chamber of Commerce president Walter Francis Timon also pushed for the deep-water port.

The city got that, too, and long before the seawall. In 1922 President Warren G. Harding approved construction of a thirty-foot-deep ship channel, which opened on September 14—note the date—1926.

So Corpus Christi recovered and rebuilt. Even those who had lost almost everything saw hope. As a businessman wrote his wife in San Antonio:

"My office is flooded and the compress completely washed away. Don't know what our loss will be but am thankful I am alive."[22]

## Chapter 13

# The Austin Tornadoes of 1922

## *"It was a terrible thing"*

**E**.J. Walthall, managing editor of the *Austin Statesman*, was ready to call it a day. It was after 5 P.M. on Thursday, May 4, 1922, but when he looked out the window in the newsroom at Seventh and Brazos, he started yelling. A tornado was coming through the city of about twenty thousand residents.[1]

Walthall and William J. Weeg began working the phones, bringing the composing room workers and others back to rework the front page. Reporters called in with news of the storm, including one who had been covering a ballgame at Clark Field. Weeg soon left the office with the circulation department's Bert Anderson, driving to the Travis Heights area in Anderson's Ford to see the damage. After a half-hour, Anderson and Weeg returned to the *Statesman*. Weeg started writing the news story as other reporters called in information. By 7:15, the *Statesman* had a new edition on the streets.

"Cyclone Takes Heavy Toll," the banner headline exclaimed.

Weeg later recalled, "It was the fastest extra ever put out in this part of the country."[2]

The *Statesman* reported:

> What appeared to be a funnel-shaped cloud was watched by crowds of people as it passed over the business district, and it had barely passed when alarming reports began to pour in from north, south and west of the city.[3]

It was the first—and only—significant tornado in Austin history.

A twister passes over the Capitol on May 4, 1922,
one of two tornadoes that struck Austin that day.
*Texas State Library & Archives Commission*

Actually, there had been two tornadoes that hot afternoon. One went through East Austin, Travis Heights, St. Edward's University, Penn Field, St. Elmo, and Manchaca, while the other passed over the State Institute for Deaf, Dumb and Blind Colored Youths, and through Deep Eddy and Oak Hill. University of Texas weather observer Fred Morris said the formation of the peculiar "double twister" had been almost perfect.

Mrs. H.H. Henderson, however, said the storm had originally had only one funnel cloud. As the funnel moved down Congress Avenue, a bolt of lightning hit and split the tornado. "One part of it struck at St. Ed's and the other at Deep Eddy," she said.[4]

Merle Wells had been selling the *Saturday Evening Post* at Seventh and Congress when the storm hit. He took cover as rain and hail fell. At Hyde Park, Mrs. Clyde Rogers watched the storm from her parents' home. "It dipped and turned and roared and I could see trees coming up," she recalled. "It was a terrible thing."[5]

John Henry Faulk and Jack Kellam probably didn't find it quite so terrible. Faulk's mother and father had left the boys at home. Faulk's father called and told the boys that a big storm was coming and they should take care. John Henry and Jack climbed into the barn and watched the tornado from the loft.

Geology professor Frederic W. Simonds had been in his home one and a half blocks north of the University of Texas campus that afternoon when he heard shouts outside. He went outside to see "a quickly advancing, dark and threatening mass of storm cloud" from which dropped a funnel that instead of hanging down, inclined off at a forty-five-degree angle. "Had the funnel hung perpendicularly, striking the ground here and there as is usually the case, the damage in and about Austin would have been much greater," Simonds noted.[6]

The storm unroofed a house near the Georgetown-Spicewood intersection, destroyed another, and pounded the Institute, where a laundry and another building were destroyed, tearing the roofs off others and blowing some from the foundations. Sheds vanished, and only the floors remained of garages. Despite the destruction, no one at the school was seriously injured.

Next, the tornado uprooted and stripped trees at Deep Eddy and damaged buildings. Southeast of Oak Hill, the storm killed six people when the tornado destroyed a home. Timbers and logs from the house vanished; about all that could be found of the house were a few stones from the chimney.

Meanwhile, the second funnel cloud had formed in east Austin and moved southwest before turning west. The twister uprooted a tree in the State Cemetery, tore the roof off the Fogle and Fogle store on East Sixth Street, knocked a house off its foundation and destroyed the front gallery, crossed the railroad tracks, and took the roof off a small Producers Oil Company building. Next, the tornado damaged a fire station and small houses and outbuildings in the Colorado River bottom before crossing the river and moving into Travis Heights.

Mike Connolly had been laying brick on a Travis Heights house when he saw the storm. He ran to his Ford, but the twister beat him. Connolly grabbed hold of a tree and held on tightly. The tornado uprooted the tree and whipped the man around. Surprisingly, both man and machine survived. Connolly was dizzy. His Ford had been set back on the ground by the twister, no worse for the wear.

Others in Travis Heights also miraculously escaped injury or death. One man who had been sitting in a tent was picked up by the wind, whirled around, and set down on the ground. In Charles Ecklund's brand-new house on a bluff, the tornado lifted the house off the ground and destroyed it. Yet Mrs. Ecklund and a friend, who had been sitting in a front room, weren't hurt.

One survivor recalled:

> Well, when that cyclone come along that hit Travis Heights, I'se out at the Woodward Body Works. I seen it coming past St. Edwards College. A red-headed feller working with me yelled, "Cyclone coming" and he broke and run plum the other side of Barton Springs. And then he seen that prong of the cyclone sucking the Colorado River dry at Deep Eddy, and broke and run back. All this three- or four-mile run in the time it took the cyclone to move maybe three hundred yards from St. Edwards College over to Woodward Body Works. . . . It was dippin' down and around, and you'd say it looked like trash. But it was really houses and trees way up in the air.[7]

More roofs and trees fell victim to the storm before the tornado hit St. Edward's, destroying the upper level of a three-story dormitory and wrecking the power plant and gymnasium. One student, caught in the open, was killed. Next the twister swept over Penn Field and hit the Woodward Manufacturing Company, killing two, injuring several, knocking a steel water tank to the ground, and wrecking one frame and four brick buildings. The eighteen-month-old daughter of Mr. and Mrs. J.R. Padgett was killed when the storm hit Penn Field. The Padgetts were also injured.

At the St. Elmo schoolhouse, the frame building was moved some twenty feet from its foundation and its roof was speared with splinters, probably from Penn Field. The storm turned west then, demolishing the Hartkoff Dairy.

Hartkoff had been milking his cows by machine when the tornado struck and decapitated twenty-four Holstein cows. The heads fell to the ground, but the farmer never found the rest of the animals' bodies.

After the storm, hailstones, some larger than a pigeon egg, fell.

Mr. and Mrs. H.H. Henderson went to the St. Edward's campus shortly after the storm. Mrs. Henderson recalled in 1954 that "you couldn't get near the place" because of the destruction and ambulances. Her husband was the chief operator for Western Union. He would stay busy over the next few days.

"All the mothers of those kids were wiring us here and the telegraph people worked all night and all day," she told the *Austin Statesman.*[8]

When the storm was over, thirteen people were dead, forty-four injured, and property damage had been estimated between $300,000 and $725,000, although the latter figure seems quite high. The dead totaled six at Oak Hill, three at Penn Field, two at Manchaca, one at St. Edward's, and one at St. Elmo. Among the wounded were Mrs. W.F. Woodman, who had a splinter driven through her head, and Jack Mussett, who suffered a fractured skull when he was pinned beneath debris and timbers at the wrecked Woodward Manufacturing Company. Houses had been "torn into splinters," streetcars were stalled and trolley lines destroyed.

But it could have been much worse.

The twisters never came down in the center of Austin. Hyde Park and downtown Austin escaped the storm's brunt.

One survivor recalled that one prong of the tornado "hit a bluff and bounced over Travis Heights or it would have wiped all of South Austin off the map. When a cyclone hits a bluff it'll bounce for a mile and a half up in the air and then come right back down as soon as it gets a chance."[9]

And yet, as with many tornadoes, there were many bizarre stories.

J. Frank Dobie reported one man's recollections thirty-six years later. When the tornado crossed the Colorado River near Deep Eddy "It sucked the Colorado River dry for a half mile each direction—so dry that a person could walk around on the bed of the river till the water ran back together."[10]

The man also recalled:

> The way it skipped some places was curious.
> There was a stack of sawdust right next to a water
> trough. The cyclone picked up that [trough], didn't
> spill a drop of water out of it, twirled it way up in the
> air and didn't even stir that sawdust setting right next
> to it. That's what a cyclone'll do.[11]

Paul Johnson had been standing in his backyard, holding his infant brother, when his mother saw the approaching storm and yelled at him to come inside. Too late! The twister passed them and sucked the baby's diaper off, leaving the infant naked but him and his big brother unhurt. The mother scolded:

"I told you not to stand out there when a cyclone was coming."[12]

And, in what seems to be a fairly common occurrence during Texas twisters, many chickens were plucked of their feathers.

The tornado was also well photographed.

Some photographs were taken from the Littlefield Building and at least one from a Congress Avenue building by Gazley Company and Jordan Company photographers. D.E. McCaskill of the University of Texas Photographic Laboratory took a series of five views of the storm, and R.L. Cannon shot two frames from a third-story window in the Main Building of the University of Texas. Many of the photographs of the actual tornado, as well as of its destruction, were printed in Simonds's account of the storm in the *University of Texas Bulletin* of February 15, 1923. The 1922 twin tornadoes might be the most thoroughly documented twister captured on film, except for the April 2, 1957, tornado that passed through Oak Cliff and West Dallas.

Immediately after the storm, however, people weren't interested in photographs or history. Many remained frightened. On Thursday night, crowds gathered along Congress Avenue to talk about the storm. Austin residents were in a "state of intense anxiety." Their anxiety was relieved, however, by an interesting source. Can you say "Hook 'em, Horns"? Around 9 P.M. hundreds of University of Texas students had a shirttail parade down Congress Avenue and other streets. The students even entered movie theaters and gave college yells.[13]

Calls for donations for the storm sufferers quickly went out by the Red Cross. The first cash donation came from the Austin Ku Klux Klan. Before the twister, the KKK had planned a parade, and Constable Charles H. Hamby had begged Governor Pat M. Neff to send the militia and/or Rangers to stop the parade. "I have no regular deputies," Hamby wrote. "Time is short . . . and I am calling on you in the first opportunity."[14]

Other donations—cash, clothing, etc.—came in from more respected sources, the Woodward Manufacturing Company and other businesses and homeowners began to reconstruct their homes, businesses, and lives. The dead were buried.

Yet by Saturday, sightseers flowed into the city. Maybe they were surprised by what they saw.

The *Austin Statesman* apparently thought so as it triumphantly reported:

> Austin was the mecca Saturday for hundreds of curious tourists from nearby towns who drove over for the purpose of viewing the havoc wrought by Thursday's tornado. Every section of the city where damage was done by the storm was visited by them. Those who expected to see dispirited individuals bemoaning their losses were greeted instead, however, with the sight of busily working people, occupied in reconstructing the damaged property left in the tornado's wake.[15]

*Chapter 14*

# A Day of Tornadoes, May 6, 1930

---

## *"Every violent gesture at nature's command"*

Texas sits on the southern end of Tornado Alley. With the Gulf of Mexico to the southeast and the Southern Rockies toward the west, the state is primed for tornadoes, especially from spring to autumn. State meteorologist George W. Bomar points out that under the right atmospheric conditions, "a whole horde" of tornadoes can be produced instead of just one. Consider the twenty-six that struck Texas in 1961 during Hurricane Carla or the more than one hundred tornadoes that hit southern Texas during Hurricane Beulah in 1967.

And then there was the outbreak of May 6, 1930.

The carnage lasted for approximately twelve hours and stretched across West Texas to deep South Texas and far East Texas. Abilene, Austin, Baird, Bronson, Bynum, Ennis, Fort Worth, Frost, Gonzales, Irene, Kenedy, Massey, Mertens, Mineral Wells, Nordheim, Runge, San Antonio, Spur, Winters, and other places suffered damage that totaled almost $2.5 million.[1]

Strong winds began that morning in Abilene, Austin, and Spur. Between noon and 9:30 P.M. the violence increased. Three Austin residents were injured that morning when the seed and hull house of the Farmers and Ginners Cotton Oil Company was destroyed. Homes were damaged, and one of the city's one hundred-foot-tall light towers on Seventh Street was blown over, the basement at the state capitol was flooded with water more than a foot deep, and traffic was stalled. A sandstorm "with a fury probably unsurpassed in this section's history" hit West Texas, leaving Sweetwater "virtually in darkness at noon" and destroying oil derricks and equipment and causing an estimated $150,000 damages in Coleman County. A Mexican man died when his house was destroyed twelve miles southwest of Spur, and four others in the house were injured.[2]

Austin took a hit from a tornado on May 6, 1930—
as did several other Texas communities.
*Austin History Center, Austin Public Library*

A piece of flying timber knocked McMurry College student Thelma Carter unconscious when the college's gymnasium was demolished. Frank Miller received leg injuries when he was blown from a boxcar at the Texas & Pacific Railway yards in Baird. An eight-year-old girl suffered a broken leg when a garage was blown over at her home near Winters. A Mineral Wells woman was injured when the roof of her apartment was blown off and the walls caved in.

Several cities recorded their highest winds that day. Dallas, fifty-seven; Del Rio, forty-two; El Paso, forty-four; Groesbeck, thirty-one; Houston, thirty-eight; Palestine, thirty; Taylor, twenty-eight. The highest, however, was at Abilene.

Winds reached sixty-six miles per hour for a minute and fifty miles per hour for five consecutive minutes just before noon at

119

Abilene, where, in addition to the McMurry gym, roofs were blown off at the Abilene Cotton Oil company and two Simmons University buildings; several barns were damaged, roofs were blown off houses, windows were smashed and trees uprooted. The sixty-six mark was a record, meteorologist W.H. Green told the *Abilene Morning News*. Several stores were damaged. Cars in the Carothers Motors Inc. showroom had to be moved to the rear of the building after glass sections were shattered. Telephone poles were blown over, and the Reverend A.C. Turner said Immanuel Baptist Church was "blown from its blocks."[3]

The storm halted a conference, with more than one hundred fifty attending, at the Lueders Baptist encampment grounds. In Ranger, two horses were killed and a blacksmith shop and other buildings were damaged by winds, small houses were blown over and roofs ripped off in Trent, and winds reached fifty-five miles per hour, taking down power and phone lines in Fort Worth. The winds were the worst in twenty-five years in Fort Worth, throwing pedestrians against downtown buildings, but damage was slight.

By midafternoon the storms escalated. Bynum, Irene, Mertens, Ennis, and Frost took the brunt of the assault, with damages reaching $2 million. The tornado missed Bynum by about two miles, but Frost suffered horribly. A pilot flew over Frost after the storm and said the tornado came from the southwest, moving northeast for about ten miles.

The day started out typically for the small town, but a dark cloud appeared in the south and another in the west. Survivors said two tornadoes struck fifteen minutes apart. "Most of the victims were trapped in their homes with no knowledge of the impending disaster," but League Wooley recognized the danger and went to the school to take his seven-year-old daughter, Mary, home. He reached home just as the storm struck. He was knocked unconscious; both wife and daughter died. The tornado took the roof off at the school and shattered every window, but no student was injured.[4]

Farmer T.J. Slay had planned on taking daughter Dorothy from the school but decided against that when the tornado struck. "After seeing it rip through the business district as if it were cardboard I knew that if the children were released from school they would all be killed by flying timbers or dashed to death by the wind." Slay said he and two teachers locked the students in the schoolhouse and "snuggled them together in the rooms and hallways." Another report said the school superintendent ordered the students to the basement. As soon as the children reached the lower floor, the roof was torn off.[5]

In the business district, more than twenty people survived when they took shelter in vaults at the First National and Citizens' State Banks. The First National's clock stopped at 3:26 P.M. when the building collapsed. The Citizens' State Bank was also destroyed.

Farmer William Meadows and his wife had been driving into town when they spotted the tornado. They jumped out of their car and threw themselves into a ditch just as a small house was blown over their heads. The car was wrecked, but the Meadows were not hurt. Another motorist had pulled into a filling station when the storm hit. He crawled under his car, which protected him when the building crumbled.

"We never knew what hit us," said Edgar Bowman, who escaped injury but saw his parents killed and his wife and son injured. "I had been looking out the window noticing the heavy black clouds, one from the west and one from the south, and they seemed to strike at the same time."

"We were all sitting in the house when it struck. The wind turned the house over on its side before we knew what had happened. Our home was demolished and all of our furniture scattered over half a block. The storm seems to have cut a swath about 30 feet wide, and our house stood squarely in the center of its path. I just can't realize that everything is gone. It was so sudden it is almost unbelievable."[6]

Fire broke out in the mangled business section, probably caused by a lightning bolt. Heavy rains put out the first fire, but another soon broke out in the residential area and had to be extinguished by volunteer firefighters.

A call came for "all available" doctors. Nearby Italy responded by sending out four ambulances and a rescue party. Relief workers coming to Frost first saw a cemetery covered with sheet iron, lumber, papers, and overturned tombstones. Ambulances had to ford swollen streams, and many stalled and had to be pushed by the drivers and rescue workers. Wooden and brick debris blocked the streets. "Automobiles were tossed about like toys. One street was almost blocked with automobiles that had been in a dealer's window."[7]

Corsicana's Salvation Army and women from the city provided coffee and sandwiches for rescue workers that night. Andrew G. Steele, county superintendent and past commander of Corsicana's American Legion post, arrived in Frost with thirty-two Legionnaires to patrol the streets and lend aid until relieved by the National Guard.

"This once thriving town resembled a war-torn village today after nature had rushed against it with its worst weapon—wind," a United Press International correspondent reported.[8]

It was a scene of chaos, the UPI report continued, as "A blasting funnel of wind and cloud descended. . . . Terror rode through the town and surrounding rural sections. Those who were not injured could talk only in unconnected phrases of what happened."

Most businesses were wrecked, and the depot "was reduced to a pile of debris and box cars were picked up and crushed by the force of the gale." The Presbyterian church survived, but the Baptist and Methodist churches were wrecked. The jail also stood.[9]

The day was far from over. Another tornado—or tornadoes—pounded San Antonio, Kenedy, Runge, and Nordheim later that afternoon.

At San Antonio, one worker was killed and another seriously injured when a storm—some said it was a tornado—struck Randolph Field between 3 and 4 P.M. Several other Randolph Field workers were injured, roofs were damaged, tents destroyed, and a construction tower blown over.

Charles Dupree, a worker at Randolph Field, described the storm that injured fifteen others:

> When the wind hit us, everyone was bewildered. I ran for the cellar, and when I got there I looked up at the first floor, and the men were all clinging to the posts, right out in the open. So I had to run back there and push them into the cellar. They were hanging on for dear life. I shouted to all of them, grabbed some of

them, pushed them into the cellar and jumped in after them. We were safe then.[10]

Hail and wind did more damage near San Antonio. Telephone poles were snapped near Alamo Heights, a garage destroyed at one residence, and beehives overturned on a farm. On Blanco Road, as Leon Horn ran from his barn to the house during the height of the storm, a twenty-five-foot-tall steel windmill toppled over—just missing him.

Nordheim and Frost "bore the brunt of violence," the *Gonzales Inquirer* reported May 8. "Every violent gesture at nature's command was exerted in one of its most vicious assaults upon the settled southwest."

About six miles separated Runge and Nordheim. The twister struck between the two towns, cutting a swath estimated at between one hundred fifty and three hundred yards wide and ten to fifteen miles long as it passed two miles north of Runge and three miles south of Nordheim, moving southeast. "I noticed a big gray cloud which seemed to be moving swiftly from the west," E.D. Parnell of Runge said. "As it approached the air grew still and sultry. The cloud came lower and grew blacker." The storm hit with destructive force. In Nordheim, the offices of P.J. Haynes and Dr. C.E. Duve became a makeshift hospital. Mexican victims—dead and seriously wounded—were taken to a Mexican hall, and many people with minor injuries were treated at homes. In Runge, one report said a vacant building (a house according to one report, an old bank in another) was turned into a hospital, the cots provided by town residents.[11]

The May 7 *San Antonio Light* described Nordheim:

Everything in the path of the storm was leveled. Strips of turf were ripped up like blankets. In some instances, stout trees were uprooted. In the potato fields over which the storm passed, potatoes were uprooted. Deep scoops were cut into the earth at different points. Potatoes which were exposed to the terrific winds were peeled as neatly as though some housewife had pared them.

Several large trees were shorn of bark, leaves, and limbs in their lower sections but still had untouched foliage at the top.

Near Runge, Paul Pierce sat down for dinner with his wife and daughter when the tornado struck, carrying away the four walls of the room where they were eating. Yet the Pierces were left seated —and not hurt.

Not everyone could be so lucky.

A piece of one-by-four lumber speared a woman through her right hip and had to be sawed off before she was taken to the makeshift hospital where doctors removed the rest of the wood. Working in Runge with eight other doctors, D.Y. Wilbern feared tetanus in addition to the serious injuries. Rocks, splinters, and other particles had cut into victims like bullets. The more seriously wounded were taken to hospitals in Cuero and Yorktown. Doctors came to Runge and Nordheim from Cuero, Yorktown, Karnes City, and Kenedy. Kenedy, about ten miles from Runge, had witnessed its own disaster.

Six miles east of Kenedy at George Tips's place, Francisco de Luna died when the tornado tore his tenant house to pieces a little before 5 P.M. Another man in the house was injured slightly. Twelve people were originally reported killed at Kenedy, but some of those may have also been on the death list at Runge, which first reports had at twenty-five men, one woman, and three children.

Tips had been riding in a pasture when he spotted the tornado. He dismounted and found shelter in a creek embankment, where he was joined by a Mexican who had been plowing a nearby field. Tips's horse was lifted into the air and dropped to the ground, dying of the injuries. Tips lost several other horses, too.

The May 8 *Kenedy Advance* described the scene:

> The trees in the Tips pasture and at other places in the wake of the storm were stripped of all foliage, and muchly resembled scenes of devastation in France following the laying down of a [barrage] in the World War.

Many of the dead and injured were taken to Runge.

Kenedy escaped the horror that had been witnessed at Frost and Runge because the tornado "skirted the northern edge of the business district." The storm blew down a tree and took off part of the roof at a mattress factory and passed over town. Several homes were destroyed east of the railroad track, and the old light and ice plant was severely damaged.

Then the storm ended.

"It was only a few minutes after the storm had passed until the sun was shining brightly and there was nothing in the elements to indicate that a tornado had raged," the *Kenedy Advance* reported. "When news of the great damage that had been sustained began to come in, persons rushed to Runge, and it is estimated that several thousand persons drove into that place Tuesday night."

Other areas also suffered that afternoon. Hail damaged several fields north of Karnes City, and the Ottine section of Gonzales was pounded by a tornado, winds estimated at sixty miles an hour, and more than one and a half inches of rain.

The storm struck Gonzales at about 4 P.M., damaging houses and barns and almost electrocuting a Mexican youth who grabbed hold of a downed wire. The worst damage came three miles west of Ottine, destroying a black Methodist church and several houses.

Sherman Clark had been plowing a field when the tornado swept down, picked up his house, and dropped it upside down on the ground. Clark's three-year-old daughter died of a fractured skull. Another daughter's leg was broken near the hip joint, but Clark's wife and three other children suffered only minor injuries.

Others residents were injured when their homes were demolished. Chickens, mules, and other livestock perished, the large hand was blown off the town clock, and wind mills and oil derricks were blown over. At Tom Rhodes's house, the garage was blown over, but "a freakish twist of the wind resulted in the car being left standing without a scratch."[12]

Darkness fell on Texas, but the storms didn't stop. A tornado killed two people in Bronson. The next morning, newspapers reported the local and statewide destruction, but because of downed power lines, reports in some places were slow to come in. A headline in the *San Antonio Light* on May 7 read: "S.A. Greeted by Smiling Skies in Wake of Storm." But in Frost, Nordheim, and elsewhere, few people probably noticed such "smiling skies."

The *Abilene Morning News* reported May 7:

> Riding on the wings of tornadic winds, death and destruction held sway over many portions of Texas Tuesday and Tuesday night, leaving a death toll of 56 which was growing hourly as more detailed reports of a series of tornadoes were assembled. Scores were injured, many seriously. Property damage was immense, and at least two small central Texas towns were virtually wiped out. The central and eastern part of Hill and the western part of Navarro counties apparently bore the brunt of the elemental barrage, 40 of the dead being counted in this area, while a little farther south, in Karnes county, another twister took a toll of at least 16 lives. At Frost, in Navarro county, the business section of which was practically laid in ruins, thirty were known dead, and others were feared buried under debris of business buildings. Included in the list were three dead at Brookins, one at Abbott, three at Bynum, two at Mertens, one at Ensign, 12 near Kenedy.

More help came. On May 7 the assistant national director of disaster relief of the Red Cross district was ordered to tornado-ravaged Texas. Meanwhile, Governor Dan Moody announced that he would not declare martial law but ordered Adjutant General R.L. Robertson to send Corsicana National Guard troops to Frost to patrol the streets and offer aid. That night the Red Cross issued an appeal for $100,000 and asked for donations of clothes and food.

By May 8 a temporary morgue was set up in the debris of the Blue Bonnet Café, where counters were used as slabs. Supplies—including four hundred loaves of bread and several tubs of coffee—arrived in Frost from Hillsboro. Dallas County's Red Cross sent one hundred cots. More victims were found. Funerals began. The search for survivors continued.

And what was known of the storm? The *Abilene Morning News* reported May 8:

> The series of tornadic disturbances hit widely separated places over an area of 281,250 square miles, and from Abilene on the west to Marshall, about 450 linear miles, and from Childress, on the north, to Runge, about 625 miles to the south.

The death toll, according to *The New Handbook of Texas*, was forty-one killed at Bynum, Irene, Mertens, Frost, and Ennis; thirty-six dead in Kenedy, Runge, and Nordheim; two dead in Bronson; and three dead in Spur, San Antonio, and Gonzales. George Bomar cites only seventy-seven deaths.

Nine members of Saragoza Garcia's family had been killed in their tenant home near Nordheim. The *San Antonio Light* described the funeral:

> There was no funeral procession.
>
> A truck was backed up to the hospital and morgue. Men silently carried homemade coffins, nailed together by the men of Nordheim, and stacked them in a truck, normally used for hauling vegetable produce.
>
> The lone survivor, pitifully brave in the face of the sudden stark flash of horror which took his wife, his brother-in-law, his mother-in-law and six children, stood with his head bowed as the unceremonious funeral took place.[13]

Garcia and his brother-in-law had been plowing when the storm approached. His brother-in-law and the children, who had been playing outside, hurried into the house. Garcia stayed outside. "A few minutes later the sky split open and a whirling compact bolt of wind struck the farm, splintering the house," the *Light* reported. A hound, the newspaper said, was keeping "a lonely vigil at the spot ever since the tragedy occurred."

More funerals—"almost wholesale burials"—were held. Farmers made house checks, maneuvering through the debris and flooded roads on horseback and in wagons. Communication lines were re-established. A week after the tornadoes, Governor Moody appealed for more funds, saying the Red Cross needed a minimum of $200,000.[14]

In Gonzales, the county's Red Cross chapter met at the Chamber of Commerce on May 15 to formulate plans to assist those suffering from the storms. Sherman Clark's daughter, who had broken her hip, the *Gonzales Inquirer* noted May 15, "has been brought to town and is receiving the proper attention, as a result of efforts of members of the executive committee." The paper went on to say that underwear for children and women and clothing for infants were needed.

Texas, however, didn't have time to regroup before more storms hit. On Monday, May 12, heavy rains caused major flooding and strong winds added to the damage. Two homes, a lumberyard, cotton gin, and a building were destroyed by winds in Aledo, and two people were hurt by glass and falling timbers. In Fort Worth, a seven-year-old boy was swept from his mother's arms and drowned when Sycamore Creek flooded. Hundreds of people fled their homes near Bryan when the Brazos River threatened them, and "Dallas streets were converted into swirling streams" and one youth was struck by lightning.

The next weekend, tornadoes and floods blasted southeastern Arkansas and North Texas. The Red River rose, forcing residents in portions of Oklahoma, Arkansas, and Texas to flee, and three Texans were killed by a tornado. On the night of May 21, wind, rain, and hail caused property and crop damage and injured at least six people in northern and northwestern Texas.

It had been a violent month, but nothing had been as destructive as May 6. The death toll for that day—whether you count seventy-seven or eighty-two—ranks behind only the one hundred fourteen killed at Waco in 1953 and Goliad in 1902.

*Chapter 15*

# The Panhandle Dust Storm of 1935

## *"It was just as black as midnight"*

They called it Black Sunday.

The dust storm, which had formed in the Dakotas, swept across the Southern Plains on April 14, 1935, Easter Sunday, marring traffic in western Kansas, eastern Colorado, and the Oklahoma and Texas panhandles. The Santa Fe Railroad was forced to detour trains in Trinidad, Colorado. And in Perryton, Texas, where dust turned day into night, the storm became among the worst in history.[1]

Texas, especially in the Panhandle, was quite familiar with these kinds of storms during the Dust Bowl. They were caused by the arrival of a polar continental air mass that generated atmospheric electricity that lifted dirt high into the air in a cold boil. Author Max Evans remembers seeing giant clouds of dust as a boy in West Texas. "People were praying," he said. "They thought it was the end of the world. But I wanted to go out and play in it." By Easter Sunday 1935, Perryton had witnessed fifty such storms in one hundred four days. On March 3 a duster blackened Amarillo "as if the sun had been blotted from the universe." B.T. Ware said of the March 3 storm: "I never saw a dust storm so severe. Nothing in my experience compares to it." Another storm, moving fifty miles an hour, surprised Borger residents. Adults and children died of "dust pneumonia."[2]

This was Texas in the Dust Bowl. It was the Great Depression, where the bottom had dropped out of the wheat crop and drought and wind ruined many farmers. Longtime Perryton resident J.T.

McLarty says, "These were hard times when a dollar looked as big then as $25 would look now. People were just barely getting by. If you could get a job paying a dollar a day, you felt pretty lucky."

The weather didn't help matters. High winds and temperatures accompanied the long drought of the 1930s, spawning these dust storms. In Amarillo in 1935, dust storms lasted nine hundred eight hours. These "black blizzards" sent clouds of dust as high as seven thousand or eight thousand feet. John Arnot described the March 3 storm for the following day's *Amarillo Daily News*: "I've seen 'em brown—but never black. They must have had a house-cleaning in Hades and let the ashes fall on Amarillo."

It was nothing to see houses in the Texas Panhandle halfway covered with dirt, or to see the breather pipe of a tractor sticking out of the dirt. Dust clouds resembled smoke from oil fires. There were red dusters from Oklahoma, yellow-tan, brown, and the terrible black ones labeled "freaks of nature."

In 1991 Dan Sell, who had been farming in the northeast corner of the Texas Panhandle during the so-called Dirty Thirties, described the Dust Bowl for his son, Richard:

> The dirt storms covered a very wide area. I didn't realize it covered as big a part of the midsection of the United States as it did. From Darrouzett, Follett, on east, they didn't have that problem. We were just here on the edge of it. West of Booker was where the worst of it was. From there on, Spearman, Perryton, all of that country, every field blew. I don't know what we would have done if we had been right in the heart of it.[3]

The worst duster in Texas history might have been the storm of April 14.

Richard Sell of Perryton calls that dust storm "the very first memories—the first conscious thoughts—the beginning of my conscious life as a person."[4]

Sell was not three years old when the storm hit the family farm one and a half miles north of Booker.

> Mother came running out of the house, my two sisters and I were playing in the dirt in front of our

home. Mother's voice, filled with fear and anxiety, screamed at us—"You children get in the house— now!" I remember the fear in her voice—it frightened me. Instead of running immediately into the house I must have stopped on the front steps—I remember watching Mother frantically running, hollering, trying to shoo or chase the chickens into the chicken house. I remember looking at the sky and seeing a great, great wall of black rolling towards us out of the sky. I next remember everything was black, I was in the house, and it seemed hard to breathe. Next and last, I remember a coal oil lamp flickering on the table. I was kneeling in a chair, and I was drawing lines and making marks in the dirt that had covered the top of the table.

Twenty-two-year-old Dan Sell and his father had butchered, scalded, and scraped a hog that day and had the hog hanging in a garage by the house. Dan Sell recalled:

About the middle of the afternoon my dad, or somebody, looked up to the north and northwest, and there were some clouds up there. When we first looked it was still a beautiful day, just as still and nice. Then I said, "Well, it looks like maybe it's gonna rain." A few minutes later we looked again, and it was moving too fast. We knew then it was something different. We realized what it was going to be, so I ran to the house and got a bed sheet to cover the hog up. . . . [Wife] Alnora began to get the baby chicks in. We saw that it was a losing battle—it was moving too fast. Dad was there. He must have been there with a wagon and team because he and Donald started home over two miles west of us on the home place. That hit them before they got home, and the only way they got home was following the fence. They just couldn't see anything. Anyway, we maybe got the baby chicks in and made a rush for the house, grabbed up the kids, and was going to go to the cellar. But we got to the door, and it hit. I had one child on each arm, Richard and

Edna Mae, and Alnora was carrying the baby, JoAnn. Anyway, it hit, and it was BLACK. It was like just as dark of a night that you ever saw. We couldn't see anything. So we had to fumble around to find the matches to light the kerosene lamp. We got a lamp lit, and we just stood there holding the kids. We were scared to death because this was something new. It hadn't ever been this bad. We gave up on going to the cellar. We figured we would just wait and see what happened. It lasted about thirty minutes.

Easter Sunday had been a beautiful spring day in the Panhandle, with temperatures in the nineties. But in Perryton around 5 P.M. people were caught unaware on the highways or at picnics, the theater, or a baseball game.

J.T. McLarty, then fifteen years old, describes seeing the storm for the first time at his house in Perryton.

It was pretty clear and the sun was shining. It was a real nice day. And then I looked north and there was a big black bank coming up. It looked just like a bank of dirt coming. Lots of people thought it was the world coming to the end, and they made a run for storm cellars. We didn't go to no storm cellars. We just looked at it and saw there wasn't no funnel clouds or nothing in it, so we just stayed there in the house.[5]

The *Amarillo Daily News* reported that the dust cloud was about eight thousand feet high. But Laura Ingalls, flying a Lockheed monoplane in an attempt to set a nonstop transcontinental record, climbed to twenty-three thousand feet and still couldn't rise above the clouds of dust. She had to make a forced landing in Alamosa, Colorado.

J.T. McLarty's brother, twenty-three-year-old Bob, was watching a baseball game at the edge of town when he saw the giant cloud of dust. He immediately started running for home but stopped to warn two uncles, who had their doors and windows open: "You better close your doors and windows. There's a dust storm a-coming."[6]

Bob McLarty crossed the railroad tracks and was less than two blocks from home when the duster overtook him. He took shelter on the south side of a house. His brother picks up the story:

> It was just as black as midnight. He stood there and kept waiting for it to clear up, and it didn't, so he got to wanting a cigarette and he rolled a cigarette and put it in his mouth. He lit a match and sat there holding it but couldn't see the match, and the first thing he knows the match is burning his fingers.

Bob McLarty waited until the storm cleared a little, then tried to walk home. His mother had put a kerosene lamp in the window as a beacon. Bob saw the light and made it home safely.

Across the street, Leonard Overton had come to John Williams's house to check on his parents, not knowing the Overtons had gone to a storm cellar and the Williamses were in the basement. Overton kept knocking, then he tried to cross the street to the McLarty's house, using the kerosene lamp in the window to guide his way. But once he started, a black cloud of dust would get so thick, Overton couldn't see the light and he would have to back up and stand in front of the Williams's door. Finally, the Williamses came up from the basement and let Overton inside. He spent the night with them.

The storm lasted most of the night. Pushed south by fifty-mile-an-hour winds, the duster struck Dalhart at 5:15 P.M.; Boise City, Oklahoma, at 5:35; Borger at 6:15 ; Amarillo at 7:20; Wichita Falls at 9:45; and traveled south "with diminishing fury" into Mexico. In Dalhart, *Dalhart Texan* editor John L. McCarty reported that he couldn't see a lighted window three feet away. It was the thirtieth duster to strike Amarillo since the March 3 storm, and visibility was zero for twelve minutes. A few days later, Georgia witnessed a bad duster.

A Stinnett woman called Sheriff Bill Adams in Amarillo to warn him of the storm: "I feel that you people of Amarillo should know of the terrible dust storm which has struck here and probably will hit Amarillo. I am sitting in my room and I cannot see the telephone."[7]

When April 15 dawned, the wind blew, but without dirt in the air. At the Sell family farm near Booker, the Sells were lucky. They lost no chickens. Nor did they have to throw out the freshly butchered hog. Some thirty years later, however, the homestead was destroyed by a tornado.

But in Perryton, "there was so much dust in the air you couldn't see the sun hardly," J.T. McLarty said. "It looked just like a little round spot." A haze spread east and west across the Panhandle.

It took a couple of days for the sky to clear. Dust covered cars, streets, and homes. Houses and businesses were filled with dirt and silt. Worst of all, the storm "put the finishing touch of destruction" on the wheat crop.[8]

There would be other dust storms in Perryton and Texas. But Black Sunday would stand out.

Recalled J.T. McLarty:

> We had a lot of other dust storms that were pretty bad, but I think that was about the worst dust storm we'd ever had. We had one storm back in winter of 1933-34 when we were living in the country. Mother said that was the worst one she had ever seen; it blowed grass and stuff into our house, and we had the door shut. Blades of grass landed in my bedroom.... But we were asleep when that one come up and never did see it. But, in '35, I remember that and, boy, it was a black dust storm.

## Chapter 16

# The Drought of the 1950s

### *"This thing was just going to keep on and on"*

In Texas, all droughts are measured by the one that rocked the entire state in the 1950s.

The years usually given for the prolonged dry spell are 1950-1956 or 1951-1957, but it lasted longer in some areas—parts of the state were feeling the effects in the late 1940s—and shorter in others, mostly in the eastern regions where annual rainfall is greater. From 1949 to 1951, rainfall in the state dropped forty percent.[1]

The effects of the drought lingered. Juliette Forchheimer Schwab, who moved to Alpine in 1941, still refuses to leave water running to wash vegetables. In 1999 longtime Sanderson resident Margaret Farley wrote:

> ... I soon realized that we are still experiencing the same weather conditions. I wondered did we ever come out of that drought. I know weather conditions were better for a period of time, but it always seems we never get enough rain for any length of time.

The drought affected more than Texas, though. Ponderosa pine forests at Bandelier National Monument died in five years. The whole Southwest suffered in the worst drought in the region since the 1500s.

Jay O'Brien worked on the Moon Ranch in New Mexico between Roswell and Portales. He recalled the dry summer of 1954:

I can remember that the cowboys kept claiming that the drought would break on the Fourth of July, as it always rained on the fourth. I took them literally and watched for the rain on the fourth. Like every other day that summer, the skies were clear blue and it was scorching hot. I didn't have to work and went swimming in a tank. That evening there were a few clouds far to the west, but no rain. If I remember correctly, we had no rain that summer and it was cloudy only one day. It did sprinkle that day, but did not even register as a trace.

During those summers, stock water and grass were equal problems. With the hot weather, there was a lack of wind. There were only two windmills on the ranch. I drew the job of running the pump jack and would sleep by the windmill to be awakened by the quiet when it ran out of gas. I would refill it; so, I could again sleep to its steady sound.

But for the most part, the terrible drought is defined mostly by what happened in Texas.

Lubbock didn't record a trace of rain in 1952. By the end of that year, the capacity of Lake Dallas was only eleven percent. Dallasites suffered through fifty-two days of one hundred-degree weather in 1953, while Corsicana hit one hundred degrees eighty-two times. The Trans-Pecos region recorded only eight inches of rainfall in 1953. A year later the annual rainfall average was eighteen inches, the worst in thirty-seven years. The only green grass found in Lubbock in 1954 was at the country club; the rest was dirt. Deer came down from the mountains into Alpine to find something to eat. Mesquite trees died. In Dallas, water from artesian wells cost more than gasoline. Water restrictions were issued in more than one thousand towns. In July 1956 Texas Farm Bureau President J. Walter Hammond telegrammed Governor Allan Shivers to request emergency federal hay for livestock producers in drought areas. By 1957 two hundred forty-four counties had been declared disaster areas.

And there were dust storms.

"It was becoming a depressing situation," Elmer Kelton remembers. "Every time one of those brown storms came roaring in out of the north, you just felt terrible. You just dreaded it. It was a sign this thing was just going to keep on and on."

In February 1953 a duster swept across Perryton from the north, blown by winds reaching seventy-five to eighty miles per hour. It reminded J.T. McLarty of the dust storms of the 1930s. "It was so black and dirty you couldn't hardly see the street lights on Main Street," he said. "They had to turn the lights on, but you couldn't hardly see them. That was our last bad dust storm we had here."

C.F. Eckhart's father, Fred, had stocked three hundred sixty-five acres in Williamson County in 1951, but it soon became obvious there wouldn't be enough water to support the animals. Two years later his livestock had been reduced to three cows, one bull, forty goats, and a horse. "It was a hell of a time," Eckhart said. Fred Eckhart held onto the lease by his fingers, paying taxes although he couldn't make a living on the land.

Deer and rabbits were in poor condition. Armadillos practically disappeared, and there were few coyotes. The worst predators were stray dogs. Surprisingly, there were raccoons "coming out of your ears," C.F. Eckhart said. "I financed many dates by shooting coons, skinning and selling them to Davy Crockett hat makers."

And the Middle Gabriel River?

Recalled C.F. Eckhart: "I never saw water in it until we came out of the drought. Since then I've seen it seven-tenths of a mile wide."

It's not that it never rained. There were many false hopes. On March 30, 1951, the *Sanderson Times* reported "Beneficial Rains Break Drouth of Many Months." But the drought hadn't even begun. By July the *Times* moaned about a "relentless heat wave." Spring rains in 1953—even flooding in parts of the state—eased the dry conditions, but the drought worsened over the following two years.[2] Said Elmer Kelton:

> There might have been one year when it rained
> more than our average, but it was an anomaly. We had
> maybe a couple of big rains, but the effects were soon
> lost. . . . It just never did rain enough at one time or

often enough so you got a carryover of the effects of one rain to the next. It would get dried out between rains; things would burn. It would rain enough to revive plants, but it never did rain enough to promote the kind of grass growth that you needed.

Kelton recalled the time he drove into a rain shower and suddenly drove out of it. He backed up his car so that the rear was being rained on and the front sat on dry pavement, then he stepped outside and snapped a photograph. "It was just like a curtain," Kelton recalled. "That was one of the derndest pictures I ever took."

The drought had its beginnings along the Rio Grande Valley in the late 1940s. In 1950 Kleberg County had recorded 18.26 inches and only 1.57 inches over the last six months. The Railroad Commission of Texas called for a special hearing for January 4, 1951, in San Antonio, and the Board of Directors of the Kingsville Chamber of Commerce urged the commission to reduce freight rates on livestock feed to the county and other drought areas in the state.

In March 1951 V.W. Lehmann sent a memorandum to Robert J. Kleberg Jr. on "Some Effects of Prolonged Drought on Game" at King Ranch. Among Lehmann's comments:

This is the first year I have ever seen geese hop up on a stock trough to drink, however, or to eat cottonseed cake put out for cattle.

Javelina suffered greatly on Santa Gertrudis. Sick or weak animals are always cut out of the pack. In normal times a lone javelina is uncommon. They have been frequent here since November.

Next to javelina, deer have perhaps declined to the greatest extent among mammals. Browse was wiped away at one sweep by the November freeze. There were no winter weeds on Santa Gertrudis and Laurelis to compensate.[3]

Cloud seeders and other "rain stimulators" were in the news again. Nobel Prize winner Irving Langmuir had led a General Electric Company team in 1946 that dropped dry ice crystals from an airplane into a cloud and produced rain. "Cloud doctors" said they

King Ranch cowboys round up cattle during a
South Texas drought in the late 1940s or early 1950s.
*King Ranch, Inc., Kingsville, Texas*

could produce more water or induce rainfall by cloud seeding.
Irving Krick, who got a Ph.D. in meteorology from California Institute of Technology, formed the Water Resources Development
Corporation in Pasadena, California, in 1950, specializing in "Sur-
veys and field operations for scientific rain induction, rain
suppression, and related water problems."[4]

At King Ranch, Robert J. Kleberg took notice in 1951, but a
year later he seemed skeptical as he wrote to another "cloud doctor," Emil A. Hanslin of Wallace E. Howell Associates in Cambridge,
Massachusetts:

> I am very much afraid that our conditions down in
> this area during drought periods are not susceptible
> on the average to the various rainmaking techniques
> and devices that are known. Last year we employed
> the Precipitation Control Company to try and break

our drought. They kept two planes and other equipment in operation for almost a year. I think we did get
a few scattered showers from this operation but the
clouds were not right during the whole period and
that is exactly what we are facing now.

I am sure it is possible under normal conditions to
somewhat increase rainfall, but under normal ranching conditions, we have sufficient rainfall to take care
of our pastures and it would be somewhat dangerous
to increase this and increase our cattle stocking ration
unless we were sure we could maintain the rainfall—this I do not believe anyone could guarantee.[5]

That was the problem in West Texas. As Elmer Kelton points
out: "If you could time the rain in this country, farmers could get by
on just twelve or fourteen inches if they could just have it all when
they wanted it. But it doesn't work like that."

The drought hit West Texas hardest.

Part of the reason might be attributed to Texans themselves.
Kelton explains:

Historically, I think one big problem we had with
this part of the country was the people who came out
here and settled first came out of wetter climates, different growing conditions, and they just assumed that
conditions here were going to be just like they were
back home, the methods they used back there were
going to work. And it took a couple of generations for
it to really soak in on people that this country had to
be treated differently.

That lesson would sink in during the seven-year drought.

In November 1952 the *Sanderson Times* reported that the
drought had "already cost us a minimum of $424,561 ... But we
haven't yet finished paying our tribute to the sun." If the drought
continued, farmers would need "above normal" financial assistance.[6]

As an agriculture reporter, Elmer Kelton had a running daily
story for seven years. He interviewed ranchers and farmers and saw
the effects of the drought on them. His father was running a ranch,

too, and Kelton had plenty of family and friends trying to ranch or farm. The drought's effect on Kelton himself was psychological. "I still had a job whether it rained or not," he said. "Now it could have very well reached the point if the drought hadn't broken where I wouldn't have had a job—and nobody else would have."

Businesses closed in San Angelo and smaller West Texas towns because of the drought. One friend of Kelton put up with the drought for three or four years before selling his farm and moving back to Missouri. The drought followed him there, however, and he wound up broke.

Cactus spines were burned off prickly pear for cattle to eat. The practice had been common in South Texas for years, but it spread to West Texas as the drought worsened. The burners worked like a World War II flame-thrower, but with a smaller flame. A worker would make two or three slow passes over the cactus, just enough to burn off the stickers. Cattle usually developed a taste for the pear, which didn't have much protein value but was a good filler and roughage and had good water content and fiber. The problem

Prickly pear is burned to help feed livestock
during a drought in the early 1900s at King Ranch.
*King Ranch, Inc., Kingsville, Texas*

was that sometimes cattle would eat prickly pear that hadn't been burned, injuring their mouths. Ranchers also put out molasses blocks or liquid molasses to supplement the pear.

Drought-stunted lambs were placed in feed pens. Others were taken off the ranges and sold along with ewes. In 1956 the drought forced sheep off the Pecos County ranges for sales two months early.

What saved many ranchers were Angora goats. During the drought, the only livestock commodity selling well was mohair. Angoras didn't require much upkeep and could survive on less grass than sheep and cattle. Others turned to hogs. One rancher came to San Angelo for scrap food from cafes and groceries to feed his hogs. It helped him keep a remnant of cattle and sheep. Others turned to poultry operations. And still others had to take outside jobs. Women, who had not worked outside the home, began to take jobs in town. Men took oilfield jobs. Oil and gas leases saved many from going out of business.

"I think the people in the most trouble were not the biggest ranches or the smallest but in-betweens that took one hundred percent of their time just to keep going," Kelton said. "They couldn't spare time for some of these other things. It took all they could to feed cattle. They didn't have time for an outside job."

The cattle market crashed in 1952, about the time Harry Truman was leaving the White House and Dwight D. Eisenhower was coming in. Ranchers complained about having Eisenhower calves following Truman cows. The calves weren't worth nearly as much as the cows.

Kelton's father, who managed the McElroy ranch, sent some cattle to Kansas, hoping it would rain before he had to bring them back to Texas. It didn't, and he was forced to sell the cattle in Kansas at a loss. Another time, after a freak rain hit Sierra Blanca, Kelton sent cattle there—and again had to sell them at a loss. "I don't know what percentage they had to cut their stocking on that range, but it was severe," Kelton said.

Most ranchers had to cut back their herds. The problem was they didn't make the cut soon enough. They kept holding out, praying for rain. By the time they gave up, the livestock market was overloaded and poor. Markets suffered severely. Thousands of cattle were being liquidated.

"I can remember a time or two when they had as many as thirty thousand sheep out here on the auction yard," Kelton said. "They had to herd them up and down the railroad tracks because they didn't even have pens for them."

Government programs didn't help much. In fact, Kelton says, "most of them probably created more problems than they solved."

Feed programs had an inflationary effect. If the government announced on Friday that it would start subsidizing certain feed, especially hay, the following week at $5 a ton, by Monday hay prices were up $10 a ton. The government was out $5, and the farmers and ranchers who bought the hay were out another five.

Simple loan programs through the FHA worked, but sometimes they encouraged people to stay when it was time to sell out. Some people eventually went broke, but the loans helped others hold on. And when the drought finally broke, most ranchers and farmers managed to pay off those FHA loans quickly.

But the worst example was when the government became directly involved. Kelton recalls when the first trainload of government-bought, drought-relief hay arrived in San Angelo, a scene he used in his novel *The Time It Never Rained*. The hay was spoiled, not even worth unloading. "But," Kelton said, "somebody had managed to unload it on the government."

The drought hurt more than farmers, ranchers, and agriculture entrepreneurs.

Water rationing in cities became voluntary, then mandatory. Juliette Forchheimer Schwab recalls hearing the sirens go off around noon. Alpine residents then had ten minutes to fill up pots of water. Juliette would have to bathe in a tub of inch-deep water.

Fort Stockton drilled Trinity sand wells in 1955 to develop reserve water supply. The city called for voluntary rationing by zones in the summer of 1956, although city officials said there was no need for alarm.

There was cause for concern in San Angelo in 1953.

The city ordered railroad cars to bring in water from Lake Brownwood. Lake Nasworthy was almost down to the riverbed. A big rain helped fill the lake, and the city never totally ran out of water, although the supply would get low over the next few years.

There was more water rationing. Residents couldn't water their yards. Water was allowed for household use only. Recalled Elmer Kelton:

> I don't remember how long this went on. We just totally kept our yard from burning up by having someone haul water in from a truck two or three times to soak the yard to at least save the grass roots. A lot of people drilled water wells. In parts of town it worked; in other parts you couldn't find underground water except salty water.
>
> One of my neighbors and I were going to get together and drill a well as close to the property line as we could and share it. But I talked to one of the old-time ranchers here, one of the Nasworthys, who had ranched that country up where we lived before it was broken up for housing and all, and he said they'd punched holes all over that country and never could get anything but salt water. So we never did drill, and the people who did around us never did get anything but salt. So he saved us from a bad investment.

Through it all, though, humor remained.

The *San Angelo Standard-Times* editor decided to run a contest for the best drought jokes. He told Kelton that they'd give maybe $25 for first place, perhaps $50 in prizes total, then Kelton would send the jokes to *Reader's Digest* and the two newspapermen would split the money.

Kelton: "I thought, sure, yeah, naturally. Anyway, we had the contest, and I took the stories in with no hope of anything coming out of it. I was just trying to pacify my editor. I typed them up and sent them to my agent, and he didn't think much of the idea either, but he sent them over to *Reader's Digest* and derned if they didn't buy the thing. I think we split about $850 between the two of us, far, far more than the paper gave prize money for."

Among the jokes:

A few farmers bring their puny little cotton crop to the gin. While sitting there waiting for cotton to be ginned out, one brings out a jug of liquor and they start passing the jug around. The more they drink, the better the cotton crop gets to looking. So they start talking about what they will do with all the money they are going to get from the crop. One says he'll buy his wife some new furniture. Another says he'll buy a new car. Finally, they ask one farmer what he's going to do. He says: "Well, I haven't quite gotten out of debt yet. Pass me that jug again."

A real estate man drives an Englishman interested in buying a West Texas ranch to a property. He can tell the man isn't too interested in the awful-looking land. Suddenly, a roadrunner appears and starts trotting along the car. The Englishman takes notice of the bird, and the real estate agent, thinking fast, says: "That's a bird of paradise." The Englishman responds: "Yes, but he's a beastly long way from home."

Then there was the old story about the 1880s drought in which a newcomer looks over the drought-stricken land and says, "This would be a fine country if it just had water." A bankrupted farmer answers, "So would hell."[7]

The jokes kept people's sanity.

In the spring of 1957, President Dwight D. Eisenhower toured through San Angelo and West Texas on a drought inspection. Shortly thereafter, the drought broke. Farmers and ranchers called it "Republican rain." The drought broke hard—"They tend to do it in a spectacular way," Elmer Kelton says—with flooding in Brackettville. Three inches of rain fell on Kingsville in one hour. Dallas had the second-wettest May in history. Other counties would be forced to apply for flood relief by June.

West Texas ranchers were in no hurry to restock. Besides, when they began to restock, the livestock market went up and they couldn't afford it. Many had learned their lessons. They realized the ranges had been overstocked. Instead of running two hundred ewes or forty or fifty cattle to a section, they understood that one hundred ewes or twenty or twenty-five cattle were enough—and that number would decrease the farther west you went. When the two-hundred-and-twenty-section McElroy ranch was sold in 1965, it held fifteen hundred head of cows. Almost forty years earlier, it had been running five to six thousand mother cows.

Elmer Kelton says of the drought:

> I think a lot of good lessons came out of it, a lot of lessons in land management. Early ranchers were not greedy, selfish or trying to spoil the land or just didn't care; they didn't know. It took two to three generations of experience to teach them that.

But West Texans remained distrustful after the drought ended. It would take one or two years before many decided the drought was over.

Elmer Kelton recalls a colleague at the *Standard-Times*, a sometime agriculture editor who even after the rains kept saying that things may look good but they weren't out of this yet.

The managing editor told Kelton: "That drought's gotten to be like an old friend to Al. He just won't give it up."

## Chapter 17

# The San Angelo and Waco Tornadoes of 1953

## *"The monster from the sky"*

Novelist Elmer Kelton, then an agriculture reporter for the *San Angelo Standard-Times*, saw the dark clouds on the afternoon of Monday, May 11, 1953, and thought the city might be in for a bad thunderstorm or sandstorm. As he drove back to the office from the fairgrounds, he waved at a sheep trader. Back at the newspaper office, Kelton arrived in time to hear the police radio reports of injuries from a tornado. Kelton was asked to go to Shannon Hospital. Once there, he waited for the ambulance to arrive and was surprised to recognize the victim on the gurney. "It was that sheep trader, who had a broken leg, I think," Kelton recalled. "He had just gotten out of his car when the tornado hit him before he could take cover."[1]

The tornado—the first major twister in San Angelo history—struck the Lakeview district in the northwest section of San Angelo at 2:15 P.M. Within minutes the tornado had cut a swath five blocks wide and two miles long, demolishing all fairground structures—including the new livestock exposition building and nine-thousand-seat grandstand—leaving seventeen hundred people homeless, destroying three hundred and twenty homes, and badly damaging another hundred and eleven houses. One hundred forty-three cars and twenty-nine trucks were smashed.

Warnings had been issued by the local Weather Bureau since Sunday afternoon and continued throughout Monday. A little after noon Monday, two Department of Public Safety officers spotted a funnel over Sterling City, forty-four miles northwest of San Angelo, and radioed in a warning. They followed the cloud into the North Concho River valley as it moved toward San Angelo at speeds between ten and fifteen miles per hour. While this was going on, the highway patrol used the *Standard-Times* telephone trunk-lines to warn residents. Teachers at the combined elementary and high school at Lakeview quickly herded one thousand children into the halls, an act that probably saved many lives as the twister crushed part of the building and injured several people but no more than twelve students. One seventh-grader told The Associated Press that the wind and rain drowned out the screams and cries at the school.

Despite the warnings, few people in the Lakeview district saw the funnel before it struck, partly because strong winds had raised dust that limited visibility that afternoon. At the fairgrounds, Roy Mason spotted the tornado and took shelter in the grandstands. As the structure began to shake and the wind tore boards off, Mason climbed into the cab of his truck just before the stands collapsed. Mason was uninjured.

As soon as the tornado ended, marble-size hail and torrential rain pounded San Angelo. Ironically, the flooding was thought to have ended the drought that began in 1951. North Concho Lake, which had been almost empty, held more than eighty-four hundred acre feet on May 13. But the drought was far from over.

Help quickly came to the city. The local National Guard unit arrived at 3:15 P.M. and assumed command. Goodfellow Air Force Base troops arrived with three ambulances, four doctors, and four nurses. By 4 P.M. the Lakeview school had been evacuated. An hour later the American Red Cross and Salvation Army were distributing clothing and blankets, and they would be supplying food before 6 P.M. By midafternoon San Angelo residents had raised more than $12,000, including a $2,000 donation from the San Angelo Police Association, to help the homeless. Sixty insurance adjusters quickly arrived, and the Armstrong Brothers Lumber Company offered to build any tornado victim a home worth $5,000 for a deposit of only $250.

As the three San Angelo hospitals filled to capacity, calls went out to Brooke General Hospital in San Antonio for blood and plasma as well as the Central Texas Regional Blood Center in Waco. By nightfall, rescue workers continued to sift through mud and debris to look for survivors and dead bodies.

In the end, damages were estimated at $150,000 to the leveled Rocket Theater, $100,000 to the fairgrounds, and between $200,000 and $500,000 to the Lakeview school. Eighty-eight homes had been slightly damaged, and nineteen small businesses had been wrecked. Property damages were put at slightly more than $3 million. Eleven people had been killed, and one hundred fifty-nine injured, sixty-six of those seriously.

But the destruction in this West Texas city would be overshadowed by another tornado that hit downtown Waco later that afternoon.

A front-page article in the afternoon *Waco Times-Herald* told the story of six migrant workers who had been killed in a Minnesota twister that Sunday. "Six Wacoans Among Dead In Midwestern Tornadoes," the headline declared.

Of course, residents of the Central Texas city had no fear of a tornado ever striking Waco. There was no reason to regard the severe weather alert that tornadoes could form around Waco or San Angelo that afternoon. In fact, the Waco Weather Bureau announced at 11 A.M. that there was no cause for concern. After all, Waco was protected from tornadoes. Long ago, Waco Indians said they would be safe from the elements on the banks of the Brazos River. Early white settlers in the 1850s believed the legend. One hundred years later, they still did.[2]

Said Waco insurance agent Carl M. Barrett:

> There was an old tale that Waco was in a bowl or a valley and the Indians said a tornado will never strike here, and a lot of people actually believed it and didn't carry windstorm insurance.[3]

That false sense of security would change at 4:40 P.M.

The sky had turned a bluish black by midafternoon, and rain fell hard. Because of the tornado warnings and watches, schools had dismissed classes early. Donald Hansard recalled the afternoon as "cloudy, overcast, very dark, real muggy, very hot, very still, very little air blowing anywhere." A monster funnel formed about eight miles outside of Waco and moved in from the southwest, moving at an estimated speed of thirty to thirty-six miles per hour. The bad weather helped in one regard by limiting traffic downtown. But it also proved fatal to many residents because the funnel was so wide and the rain so heavy, few people could see the storm approaching.

Mrs. Joel Chance had been at home when huge hailstones pelted the house. She called her husband at work downtown and asked him about the weather. He told her there was nothing to report, so she went outside to grab a hailstone to put in the refrigerator to show him. A few minutes later, he called back. "Honey," he said, "town has just blown away."

Mrs. Chance picked up her daughter, who was taking dancing lessons, saw the damage, and went to work. She served as chair of the Red Cross volunteer activities. She wouldn't go home again for three days.

Carl M. Barrett also worked with the Red Cross as vice chairman. He had been at his insurance office on Austin Avenue when the tornado hit, but the twister jumped over the building and blew out plate-glass windows. Barrett went home, where the Red Cross reached him and told him to report immediately. The R.T. Dennis and Company furniture store's five-story building at 426-28 Austin, he was told, had been destroyed. "I hardly believed it, but I jumped in the car and reported to the Red Cross, and we worked all that night doing what we could."

Wilton Lanning Jr. arrived at the old Tom Padgitt Company from school at 4 P.M. He walked outside underneath the canopy with Bob Blaylock, manager of the sporting goods department.

"Junior," Blaylock said, "put your hand out there. Tell me what you feel."

Lanning did and said the rainwater felt warm.

Blaylock shook his head. "That's what they say happens when there's going to be a tornado."

Lanning went to make some deliveries in his father's Buick. When he returned, he saw the windows had been blown out of Padgitt's store. Not only that, signs were bent over all across downtown, wires were sprawling, and people walked around in a daze. Inside the store, water stood eighteen inches deep. Lanning went home, put on his Explorer Scout uniform, and returned to downtown to help direct traffic.

The giant tornado had hit downtown, sending bricks, glass, and other debris raining. Much of downtown had been reduced to rubble. As Barrett recalled: "It was described at that time as the monster from the sky and I think that's an accurate description. . . ."

At Behrens Drug Company two and a half blocks from the Dennis furniture store, a water tank fell through five stories and killed a custodian in the basement. The twenty-two-story Amicable Building, built with steel frames and brick walls, swayed some ten inches but held. The older Dennis building, of wood construction, had no such luck.

Dennis worker Billy Watkins called his wife, Betty, and told her not to come to town because a tornado was expected. Before Betty hung up the phone, she heard glass breaking. Billy Watkins's body wasn't found for four days. Twenty-two employees were killed at the Dennis building, and nine others were injured. Nor were those outside the building safe.

Baylor University Professor Keith W. James and his wife, Helen, had been driving home when the weather presumably forced them to stop. They were waiting in their car when the Dennis building collapsed, crushing them to death under a mountain of bricks and rubble.

Ted Lucenay, an assistant in Doctor Ernest Johnson's office near the Dennis building, had come back from having a cup of coffee. The doctor's office was located on the ground floor, and a physical therapy department was upstairs. The building caved in. A falling desk killed a nurse, and Lucenay suffered a punctured lung. He was buried for three and a half hours, and for a while his wife thought he was dead. Lucenay, however, felt lucky, recalling:

> My wife, who always met me at 5 o'clock in the afternoon, had stopped that afternoon to help a neighbor catch his duck. And had it not been for that, she and the two boys would have been killed in the tornado.

At the Torrence Recreation Center behind the Dennis building, Donald Hansard, a high school senior, had gone to play pool with his best friend, Kay Sharbutt. Kay parked his car in the alley around 3:30 P.M. An hour or so later, the two friends were in the midst of another game when the lights flickered, then went out. Hansard and Sharbutt continued to play between flashes of lightning while four Hispanics squatted below the next pool table.

Sharbutt made a shot to which Hansard quipped: "That was very lucky."

Moments later, the rec center collapsed. "It was like a big clap of thunder," Hansard recalled, "like a bomb going off. . . ."

The upstairs domino hall fell in on the pool players. A joist above Hansard's head fell across him and one fell on the side of the pool table. Hansard's right foot lay on Sharbutt's left arm. Hansard lay trapped with a fractured hip and bruised foot. Hansard's hand was sticky with blood, and he couldn't see his friend, who lay only four feet away. Kay Sharbutt, seemingly invincible at six-foot-six and two hundred sixty pounds, was dead. The same beam that had pounded Hansard's foot had crushed Sharbutt's skull.

Hansard never lost consciousness. He heard screams and felt water. Water lines had burst, and air blew through an air vent on his soaked clothes. He wet his fingers to moisten his mouth, picked up a pool cue, and started tapping.

> A lot of people were cursing God and didn't know what was wrong. A lot of people were begging for their mother and it kind of struck funny to me that these big, tough guys, they wanted to talk to their mother. Of course, the screams and the pleas of those people who were in there with me—and I'm sure that mine were the same; I just don't remember that much.

Fourteen people had been killed in the rec center. It would be hours before somebody found Hansard.

At a dry-cleaning and laundry-machinery business at Second and Mary, owner Roger N. Conger and assistant sales manager David Lepar looked out a plate-glass window just as the tornado hit. It "sounded like a fleet of B-29 bombers coming right over the rooftop," Conger recalled. Windows and doors began to blow in, and a bookkeeper fainted. Outside, Waco Hardware trucks that had been backed up against the dock "rolled like tin cans over and over up Second Street."

George Hutson, a local manager for Southwestern Bell, had been talking on the telephone, when a billboard blew across the street and crashed through the Safeway's sign. "Something is happening bad and I've got to quit talking to you," he said, and hung up. Outside, a man was beating on the locked front door.

"We're closed," Hutson said, but he let the man in anyway. "The whole town is blowing apart," the man said.

He wasn't exaggerating. In the business district, the Amicable building, Roosevelt Hotel, and City Hall stood, but much of downtown had been destroyed.

Several people were at the Joy Theater, between Chris's Café and the Dennis building, watching the Robert Mitchum rodeo Western *The Lusty Men* (the other feature was *Follow the Leader*). When the Dennis building collapsed, the theater was destroyed. The roof of the theater collapsed but miraculously caught and held several feet from the floor, saving many lives.

Rescuers tried to question one boy pulled out from the Joy seats, but he proved uncooperative. Finally he blurted out: "For God's sake, let me go. When I get home they're going to beat the hell out of me anyway because I was supposed to be at home working."

Response around the city was quick.

With some sixty-three hundred telephones in Waco now out of order, George Hutson left his Southwestern Bell office in a mobile-equipped telephone car, parked on a sidewalk, and put out appeals for help. He called the media, construction companies, and the military, using Sheriff C.C. Maxey's name as authority.

The state police blocked off all highways to Waco. Funeral homes converted hearses into ambulances. Citizens took shovels from the Montgomery Ward store to help dig through debris. Major William Stevenson of the Salvation Army ordered three hundred loaves of bread from the bakery and started getting sandwiches ready. Then he got on the radio and requested coffee, sugar, and milk. Doctor H. Joe Jaworski, who had been on his way with his wife to their suite at the Roosevelt, treated some people on site, then took his car to Hillcrest Hospital, a drive that took about a half hour. Jaworski still didn't realize a tornado had hit, but casualties began arriving soon. He had served as director of surgical services at McCloskey General Hospital during World War II, and that experience would help now. Jaworski worked until late the next morning without stopping.

In a hospital surgery room, a man knelt over his wife to hear her whisper, "I love you." Elsewhere, an elderly man, on a cot in the corridor, said: "Oh, God, I wish that doctor would come stop this pain." Psychologists and psychiatrists talked to people in shock. Most of the injuries were fractures. Others were treated for shock, multiple cuts, and internal bleeding. Only one person died after being received at Waco hospitals.[4]

Forty blocks downtown were closed to the general public and placed under armed guards. And rescuers began digging out survivors—and corpses—as loudspeakers on trucks blared: "Do not smoke or light a match. Gas mains are broken. . . . Don't step on any wires. Some of them are hot."[5]

Around 9 P.M. Waco National Guardsman Claude Kincannon cut a hole above Donald Hansard's head at the demolished recreation center. A Waco High School graduate and now a senior at Baylor, Kincannon owned a service station at Twenty-fifth and Franklin and had been repairing a flat at Cameron when the tornado struck, blowing the car off the jack and Kincannon into the gutter.

Kincannon saw bricks smash windshields, so he locked up his service station and went to the National Guard Armory, put on fatigues and army boots, and started digging out bodies. Kincannon recalled the nightmarish scene:

> The main thing was cars—cars almost flattened in the street, bricks all over them; that left a real image with me after seeing the bricks through the windshields come by the station. And just the complete destruction—some buildings standing. And it was frightening—you didn't now whether—how long they were going to stand.

At the rec center, Kincannon heard Hansard's taps. Other volunteers showed up with backpack shovels and picks, and they went to work. Hansard yelled, but Kincannon told him not to holler but continue tapping with the pool cue. They would dig to the noise. Darkness fell, forcing the rescuers to use flashlights. They pulled out a few dead bodies, including Kay Sharbutt's, and began trying to extricate Hansard.

Hansard reached out and grabbed Kincannon's gloved hands. Kincannon asked for the teen's name and address while others tried to jack up the beams and Phil Hartberger and Doug Huspeth tunneled under. They could smell natural gas from broken lines, so Kincannon kept Hansard talking so he wouldn't fall asleep. Shortly afterward, Hansard's father, who had originally been told his son had been killed, arrived from suburban Bellmead.

"How are you doing?" his father asked.

"Well, I'm fine," Hansard replied. "I'll be out in a few minutes."

He was in shock, but a few minutes later, a Presbyterian minister, who weighed only one hundred twenty-five pounds, pulled the two-hundred-pound Hansard out with a blanket.

Hansard went to the hospital, and Kincannon went back to work. He worked for four days before going home. There, the National Guard reached him, finally, and asked where had he been. "I never did go to roll call," Kincannon said. "Didn't know they had one. I just went to my locker and put my clothes up."

The most documented—and probably the most heroic—rescue, however, had to be that of Miss Lillie Matkin.

Matkin had been the switchboard operator at the Dennis company for thirty-three years. Matkin had just finished reminding department heads of an upcoming conference call when the building collapsed, trapping the slightly built, elderly woman behind a divan in the building. She was on her side with her legs doubled under her stomach, right hand free, but left arm pinned behind her back.

Matkin could hear people talking, and she screamed for help but got no response. Then she started praying, asking God to take her if she couldn't get free. She was ready. Rescuers, however, heard her cries around 5:30 P.M. and began digging. Moving the debris became hazardous. An attempt to tunnel through the debris on Fifth Street failed. A support was accidentally dislodged, causing a heavy chunk of debris to press against the trapped woman. Gas and water leaked from broken lines.

When she could see the rescue workers, she asked for ice, but threw up white froth. She later took food from glass or straw, and was given oxygen. Shortly before 7 A.M. Tuesday, Lillie Matkin was pulled from the Dennis building wreckage to cheers from onlookers. She thought it was 7 P.M., not 7 A.M., and thanked God for getting her out. She also asked workers not to lose her shoes.

Rescue worker Ray Batemon said he felt like crying once Miss Matkin had been freed. He added that only God knew how she had survived.

One hundred fourteen people had been killed. Another man, forty-three-year-old C.J. Ott, died of a heart attack on Tuesday morning after spending more than half the night in rescue and cleanup operations. But oddly enough, there were no other natural deaths for four days after the tornado.

Bodies continued to be found. Mr. and Mrs. John C. Neely were found in the basement of their paint store on Franklin at 2:45 P.M. Tuesday. Their son-in-law, twenty-six-year-old John W. Coates was found the next day. Coates's wife was treated for shock at Providence Hospital.

The smell of death permeated Waco. The Wilkirson-Hatch Funeral Home had to covert a garage into a temporary morgue. A department store sent bolts of cloth to cover bodies, and the Dallas morticians school and other Dallas funeral homes sent crews to help. Some bodies were identified by jewelry and clothing. Double and triple funerals were common. The last funeral was held at 9 A.M. May 18.

Limited power was restored on May 14, and military operations ended on Friday, May 15.

President Dwight D. Eisenhower declared Waco a major disaster area on May 16.

Between May 11 and 16, 9.59 inches of rain fell on Waco, but it didn't rain May 17, and the clear skies brought out sightseers, some from as far away as two hundred fifty miles, to see the damage.

Waco had lost much. One thousand ninety-seven people had been injured, two thousand automobiles and eight hundred fifty

One hundred and fourteen people were killed
when a rare tornado ravaged Waco in 1953.
*The Texas Collection, Baylor University*

Waco residents check out the damage shortly after a
tornado hit downtown on the afternoon of May 11, 1953.
*The Texas Collection, Baylor University*

Looking north on Third and Franklin in Waco, after the deadly
1953 tornado. Note the National Guard soldier in the center.
*The Texas Collection, Baylor University*

homes destroyed or damaged, and one hundred ninety-six buildings destroyed or forced to be demolished and three hundred
seventy-six buildings called unsafe. Total property damage: $51
million. Crop damages: $1.65 million.

Katy Park, home of the minor-league Waco Pirates, had been
destroyed. The season had opened April 14, but the Pirates had to
move to Longview to complete the season. One fan told the *Waco
Times-Herald*: "Worst thing about it, I'll have to stay home nights
this summer."[6]

But there were positives.

City attorney Lyndon Olson Sr. praised Mayor Ralph Wolf and
City Manager Jack Jeffrey for taking charge during the disaster,
making quick decisions. Olson also told this story about Mayor
Wolf:

> Bill Foster, used to be editor of the *Citizen* here,
> *Waco Citizen*, was terribly upset because they'd cut the
> gas off to his place of business and he couldn't print
> his paper and he wanted to print the paper. He went to
> see Ralph Wolf and Ralph told him they couldn't turn
> it on. And old Bill is a persistent sort of fellow and he

kept pushing and pushing and finally Ralph hit him, knocked him down. I remember one day he came to my office and said, "What in the world could I do? I just knocked the hell out of Bill Foster." Ralph Wolf says. I said, "Well, just help him up." So he did and nothing ever came out of that. . . .

Ten years later, Lillie Matkin said she no longer gave May 11 much thought. Dennis Company survivors filed for workmen's compensation payments in federal court, but the first trial was declared a mistrial because of a deadlocked jury and the second trial ended in a verdict against the survivors when the jury decided that the insurance companies couldn't be held responsible for an act of God. When the Waco-McLennan County Library began interviewing survivors in the 1980s, Florence Watkins said she thinks about her son Billy, killed at the Dennis building, every day.

But Waco moved forward. Said Carl M. Barrett:

It was a terrible experience but the good things that came out was the tremendous help that came throughout the nation; the way the insurance companies handled their responsibilities, the Red Cross and other organizations; people banded together, and it was a tremendous display of man helping man as far as I was concerned.

## Chapter 18

# The Sanderson Flood of 1965

### "Your town is washing away"

In 1969 the U.S. Soil Conservation Service recommended that a series of eleven dams be built to moderate the runoff through tributaries to Sanderson Canyon west of Sanderson. Estimated cost: $4,566,350. Actual cost upon completion of the network in 1987: $34 million. Recalls Sanderson native Susan Corbett: "The project was lampooned as the biggest waste of taxpayer dollars in years: Dams to hold back the tides in the desert, where rain is as rare as snowfall."[1]

A stranger driving down U.S. Highway 90 through this small West Texas town about one hundred miles northwest of Del Rio in the Big Bend country might agree. It's hard to believe that a flash flood once tore through this town, killing twenty-six people and making the *CBS Evening News* with Walter Cronkite. But for those who lived in Sanderson on June 11, 1965, the nightmare of the flood is all too real.

Originally named Strawbridge, Sanderson got its start in the 1880s when the Texas and New Orleans Railroad reached the town site. When Terrell County was established in 1905, Sanderson became the county seat. The town remained an agriculture and railroad community. By 1965 the population had reached two thousand three hundred and fifty, and the Southern Pacific Railroad had about one hundred and fifty employees, running a change and repair operation there and serving as a home for the eight train crews on the Sanderson-El Paso route.

Heavy rain began falling on Thursday afternoon, June 10, 1965—which seemed quite welcome in a county with a mean annual precipitation of 12.8 inches and a town that averages 1.83 inches in June. The rains, however, kept falling—as much as eight inches in two hours—and seldom let up. West of town, ranchers measured nine and a half inches of rain. In town, almost six inches fell between Thursday afternoon and Friday morning. An early morning deluge dumped five and a half inches of rain in Three Mile Draw between 5 A.M. and 8 A.M. All night, county and town officials watched the rising water in Sanderson Creek, but when the rain slackened around 5 A.M., many believed the creek had crested and went home to rest.

Southern Pacific conductor Thomas Corbett had been called to work on Thursday. Upon arriving at the depot, he called his wife, Frances, and said, "It's raining hard west of Sanderson. Keep an eye on the creek." The Corbetts—Thomas, Frances, and children Tommy Junior, William, Susan, Mary, and Mickey—had moved into the new house just a year earlier.[2]

Sanderson Creek had flooded before—including 1933, 1935, and 1937—but accepting a major flash flood seemed hard for some longtime residents. The Sanderson Creek watershed, however, had filled to unprecedented levels. The flooded Three-Mile Draw converged with the flooded Sanderson Creek and sent a wall of water, fifteen feet high and measured at seventy-six thousand, four hundred cubic feet per second—some estimates were as high as one hundred thousand cubic feet per second—onto the unsuspecting town around dawn.

Between 6 and 7 A.M. Border Patrolman Kenneth Epperson, who lived near the railroad tracks on the west side of town, abandoned his house and took his wife and children to a café. He quickly called Sheriff Bill Cooksey.

"Sheriff," he said, "you better get up, your town is washing away."[3]

The *Fort Stockton Pioneer* reported that the late rainfall dumped as much as twenty inches in the area, knocking out telephone lines and making roads impassable so that any hopes of warning Sanderson vanished.[4]

By 6 A.M. water was "almost touching" the bridge west of town, and heavy rains resumed. Motel managers were warned to expect water in the rooms and were advised to get people out. Others, realizing the danger of a flash flood, began warning residents in the flood zone.

Around 7 A.M. Frances Corbett received a phone call from a neighbor of her mother. "Get out of the house," she was warned. "The creek on Main Street is running full and the water is lapping over the bridge." Frances Corbett looked out a window to see the yard covered with water. She immediately woke up the five children and told them to get dressed quickly. The children didn't understand. When it rained, the creek ran so why did they have to leave the house this time. Frances wrapped the youngest child in a blanket, and the family piled into a station wagon.[5]

Daughter Susan, just a few days from her fifth birthday, kept asking her mother, "What's a flood?" Frances didn't reply. Frances had just learned to drive a few months earlier. She drove one block to the corner and stopped at the Club Café on U.S. 90.

"There I saw the wall of water coming our way," she recalled. "I had nowhere to go. . . ."[6]

Electric clocks stopped at 7:05 A.M. At that time, the wall of water swept through Sanderson, destroying buildings and automobiles and washing away homes. In forty-five minutes, half of the town had been destroyed.

Telephone poles were driven through the steel-reinforced walls of the Sanderson Wool Commission Company, washing away sacks of wool and mohair. Three one-thousand-gallon gas tanks from a Texaco station were washed downstream. Kerr Mercantile warehouses were destroyed, and the lumber yard disappeared. At the Southern Pacific yard, employees were stranded as water reached window level and twisted miles of rail. Those on the top of the depot heard cries of help over the roar of water. Fifty-eight-year-old porter A.F. Scott stood on top of an old passenger car, converted into an apartment, that was being carried down the creek.

Terrell County judge R.S. Wilkinson and his wife climbed on top of their one-story stucco and frame house. It was 7:15 A.M. Houses that had stood to the northeast by the creek had gone.

Then the Wilkinsons saw the stranded Mr. Scott floating on top of the passenger car only fifty feet away. Recalled Judge Wilkinson:

> We hollered and hollered at him. But the noise. There was so much noise he didn't hear us. We were trying to get him to grab hold of the telephone wires, but he didn't hear us.... If he had hung onto those telephone wires, he'd be here today.[7]

The body of A.F. Scott was found six days later near Eagle Pass, two hundred miles downstream.

Meanwhile, Frances Corbett sat in the station wagon with her young children, watching a wall of water rapidly approach. Suddenly Francisco and Marta (or Martha) Lopez, with their children Jamie and Thomas, drove past in a car driven by Alfredo Calzada. Marta yelled at Corbett to drive south toward Dryden some twenty miles away. Frances pulled out in the wagon, praying out loud. Susan asked her mother who was she talking to. And the children, pretending this were a game, shouted: "Hurry, Mom, the water is catching up with us!"[8]

Frances Corbett recalled:

> I went as fast as I could and as we got to the first bridge south of Sanderson the water was going over the bridge. We got up to the hill and I kept going 'cause I could see stuff floating everywhere.[9]

Susan Corbett said: "It was an adventure for us children, a hellish nightmare for Mother."

The Corbetts came to another small bridge, where several trucks and cars were stranded. Half of the bridge had been washed away, but a truck driver asked Frances if he could drive the station wagon across the bridge. He needed to call his boss, and his rig wouldn't make it over the bridge. Frances agreed, the station wagon reached Dryden, and the truck driver bought the children breakfast. The power went out and the store owner, a Mr. Ten Eyck, announced he was closing the store and going to Sanderson. The Corbetts climbed back into the station wagon and followed the Ten Eyck family. They reached the Sanderson bridge, which had collapsed. It was no later than 11 A.M. "We stayed there for a long time," Frances Corbett recalled.[10]

Susan Corbett's grandmother, Natividad Venegas, was having an even harder time. Swept off her feet by the churning waters, she prayed. Meanwhile, Augustine Maldonado had just rescued a boy trying to flee the floodwaters when he saw Venegas being swept down the road. Maldonado stopped his highway department wrecker, waded through the water, and pulled her to safety. Maldonado and Polo Calzada got Venegas to higher ground and took her to Calzada's house one block from Venegas's house.

Maldonado also rescued two other women gripping a fence near the bridge. Maldonado and Cuthberto Gonzales brought the two women to safety, then the two men rescued other women trapped in their houses. Maldonado's wrecker, however, was soon swept off the bridge and found, totaled, two miles downstream.

Others were not as fortunate.

The family of John Wesley Johnson had been staying at the Robertson Courts Motel when the crest slammed into the buildings. Railroad brakeman Morris Nichols had been asleep in the motel when the water knocked his door open. He pulled on his pants and went to the Johnson's cabin. The twenty-two-year-old Nichols told The Associated Press:

> I hollered at John "Let's get your kids out of here. There's a flood coming." I told him to give me three kids and I'd try to make it to high ground. He said he couldn't open the door, the water was hitting the door so hard, and he told me to go around back. The water was chest deep then.
>
> I walked or swam around the cabin and was in the garage when the big wall hit. I was trapped. I finally got a hold of the door and pulled myself free from the garage. The water was over my head then.
>
> I could see the water rushing out there, turning over cars like they were toys. I tried to get on the roof of the motel but I couldn't pull myself up. Finally I turned loose. I thought I was going to die.
>
> It washed me about 300 yards, turning me over and over. I caught a hold of a telephone pole and straddled it. A car washed in and pinned me to the pole.

I looked back and saw all of the Johnson kids up on top of the motel. John had put them up on the roof through a back window.

The motel started crumbling and it went over and everybody was gone. I knew the kids would die and I couldn't help them. Nobody could help them.

Then the water started splashing over my head and I turned loose of the pole. I went about a quarter of a mile and I was able to touch ground.

I found two of the Johnson children. They were both dead. I was completely dazed. I was hurt, cut, and beat. I found the highest place and sat down on a hill and prayed.[11]

The body of John Wesley Johnson, thirty-five years old, was never found. Said Nichols: "Their daddy was a big strong man. He might have made it but he probably tried to save his kids and was lost." Also dead were thirty-one-year-old Ethel Johnson, fourteen-year-old David Neal Johnson, ten-year-old Lou Ann Johnson, nine-year-old Jo Ann Johnson, eight-year-old Rose Ann Johnson, and five-year-old Paul Daniel Johnson. The family had come to Sanderson from Fort Stockton for temporary construction work.[12]

Only one family member, twelve-year-old Michael Johnson, survived. Transported by Air Force helicopter to Pecos Memorial Hospital in Fort Stockton, where he was treated for exposure and minor injuries, Michael told the *San Angelo Standard-Times*: "I saw the water coming. I grabbed a tree. Mom and dad and the children were yelling and screaming and I saw the wall of water go over them.

"If we don't hear from mom and dad by night, we know they are dead."[13]

Also at the motel, the Sellers family (in town on a construction job) had been surprised when water rose to the level of their beds. They placed their children, Jerry and Frankie, both six, and Debra and Kenneth, both five, in a tree outside the window. The children were swept downstream, however, and drowned. The parents survived.

As Morris Nichols told The Associated Press:

> As far as I know there was not another person in that motel alive. You can't imagine what happened when that wall of water hit. People were drowning and you just couldn't do anything for them.
>
> Women were looking for their little children. You know they're gone but there's not a damn thing you can do. When I was on that pole I saw a house come down the street with three people on it. It just turned over and I don't know what happened to them.[14]

Thirty-two-year-old David Flores also survived a horrific ordeal.

Flores had waded through rising waters to the home of his elderly grandparents on South Fourth Street when the crest hit. He helped his grandparents onto chairs and then a table as the water rose. When the water neared the ceiling, Flores knocked out the ceiling, held onto the ceiling joist with one arm, his grandfather with the other, and his grandmother with his legs. When part of the west wall caved in, the current took his grandmother under. The rest of the wall soon collapsed, and his grandfather was swept away. When the water level finally dropped, Flores freed himself and looked out the window. He saw his grandmother gripping a thin cypress tree in the front yard.

"There she was," he told the *San Angelo Standard-Times*. "Nothing but the grace of God saved my grandmother."[15]

His grandfather, seventy-year-old Nicholas, however, had drowned.[16]

David Flores carried his grandmother to higher ground. He saw thirty-six-year-old Jesus Marquez at a service station on U.S. 90. Marquez tried to throw a wrecker cable to Flores, but the current proved too swift. Flores waited there until the waters receded. Marquez must have left to try to save his own family. "That was the last time I saw Jesus," Flores said.[17]

Marquez drove his wrecker toward his own home, but the high water soon killed the engine. He continued on foot in waist-high water and finally climbed on a shed near the house where his wife, Bertha, and two daughters were trapped. Bertha saw her husband on top of the shed. He told her to stay inside. As the water rose and walls began to crumble, Bertha thought the house was about the collapse so she went outside with her two children, trying to reach some cars about ten yards away. A wave swept Belinda, the oldest girl, away. Bertha couldn't find the girl and ran back inside the house, praying for God to take Belinda soon. The girl survived, however, and made it back inside, naked and blue. Bertha looked outside for her husband, but Jesus was no longer on the shed. Apparently, he had jumped into the water—although he could not swim—to rescue his family. His body was never recovered.

Meanwhile, Sheriff Bill Cooksey, Jesus's brother Steve, and other men had arrived on the scene. They secured a rope to a car and tree and inched their way toward the Marquez house. Joe Fuentes carried Belinda to safely, and Bertha and Rosella were also rescued.

Albert "Beto" Escamilla had loaded his family in his truck and deposited them safely on higher ground. Then he drove back to Sanderson to aid in rescue efforts. One of the first persons he found was County Commissioner Frank Wiggins, whose car had been washed off the road. Escamilla spotted Wiggins hanging on a telephone pole. Escamilla got out of his truck, swam to Wiggins, and helped the commissioner to safety. Escamilla later drove to the Marquez house and saw Jesus Marquez trying to rescue his family. Escamilla found a rope and tried to toss it to Marquez. Later, Escamilla recalled:

> People were so scared they didn't know what was going on. I was trying to throw [the rope] to him. There wasn't anything I could do. I knew he was going to be washed away and I just didn't want to watch it.[18]

The flood also rushed over the cemetery, unearthing coffins. As the waters receded, the survivors were shocked by a new horror.

Frances Corbett sat in the station wagon with her children when a highway patrolman approached her and asked if she could help identify bodies that had been discovered.

> Some were bodies that had been washed out of the cemetery. They had a crucifix in their hands. One of the bodies was that of Francisco Lopez, the man

whose wife had told us to go south. He was all swollen and bruised. His shirt was gone. All he had was the collar of his shirt and a little bit of his sleeve.[19]

The Lopez family and Alfredo Calzada had been stranded when a tire blew out as they tried to make it past the last bridge. Calzada, thinking it safe, had pulled off to change the flat when a wall of water surprised him and the Lopezes. Calzada tried to grab hold of one of the boys, but the current swept the entire family into the canyon and to their deaths. Calzada himself was swept about half a mile down the rapids until a butane tank caught in the roaring waters somehow lifted him to shore.

By the time many Texans were just sitting down to breakfast, the floodwaters were receding.

"When it was all over, I was in a daze," Morris Nichols told The Associated Press. "I just walked and walked. It was like a dream—a bad dream seeing those children dying like that and you can't do anything about it."[20]

But the danger wasn't over. Sanderson—what was left of it, anyway—was cut off. The roads were washed out, and all phone lines were down.

Border Patrolman Kenneth Epperson suggested that an airplane could be flown to Fort Stockton. The plane was towed to Highway 35, where there was room to take off, but the plane wouldn't start. Sheriff Bill Cooksey saw a dazed man walking down the street with a dead boy in his arms. "I realized then, I think for the first time, how serious this was," he later told the *San Angelo Standard-Times*. "It had happened so fast no one had time to think."[21]

Cooksey started an ambulance and went down Main Street. Railroad commissioner Philip Hansen was taken to the home of Jolly Harkins, who owned a fifteen-hundred-watt transmitter. Kenneth Moses provided the generator, and Hansen began sending out distress messages on the short-wave radio. The receiver didn't work, so they had no idea if their call was being heard, but Hansen kept repeating: "May-day, May-day. . . . Sanderson has been hit by a flash flood. . . . May-day, May-day. . . . Sanderson has been washed away. . . . We need assistance."[22]

The report was heard in La Mesa, New Mexico, and telephoned in to Pecos, Texas.

Help came quickly. Alpine and Fort Stockton radio stations began appealing for donations of food and clothing. By late Friday afternoon, a refrigerated truck had arrived. Doctor John Pate arrived from Alpine. The Red Cross in Washington, D.C., sent word to the Fort Stockton chapter. Fort Sam Houston in San Antonio supplied Huey helicopters. The highway patrol set up roadblocks; the Border Patrol, Salvation Army, and neighboring county sheriff's departments sent help. A Fort Stockton milkman brought milk and baby formula, a water truck arrived, and local merchants provided food and other supplies.

Sanderson needed help.

Many businesses were ruined. Fifty-four homes had been destroyed. One hundred and sixty-nine had been damaged. Four hundred and fifty people lacked shelter. Losses exceeded $1,680,000. The elementary school was temporarily turned into a shelter.

Rancher Eddie McNutt told the *El Paso Times* he couldn't get to his home because of mud and high water and wondered if anything remained.

Conductor Thomas Corbett had even more reason to fear.

Corbett had been sent to Snyder. He and the crew walked into a restaurant on Saturday morning where they saw a newspaper headline about the Sanderson flood. "My God, my family were in the flood!" he yelled. "I have to go home." Corbett took a bus and arrived in Sanderson on Sunday afternoon. He found his family—originally on the missing list—and told his wife, "Frances, I've seen the house from the road. It's still there." Frances Corbett, however, had already seen the house.[23]

Frances and the children had been taken to the elementary school. They stayed at an apartment until their house was repaired. Frances Corbett recalled:

> Our house was still standing, but the inside was full of mud and everything ruined. We found the furniture piled around in different rooms. The mud came up to my knees. We were asked to leave 'cause snakes were everywhere.[24]

It was like that all over Sanderson.

"This is the damndest mess I've ever seen," County Commissioner Sid Harkins told the *San Angelo Standard-Times*. "We've got a real job on our hands getting it back in shape, but we can do it."[25]

Sheriff Bill Cooksey reiterated Harkins's statement:

> People who live in West Texas get used to nearly every kind of disaster. This is a great disaster to our community but I don't think that it will put us out of business. As soon as the water runs down, we'll be rebuilding.[26]

Electricity and telephone service was soon restored, but as the rainfall ended, temperatures climbed to more than one hundred degrees. Health officials feared the unearthed bodies, silt, and ruptured cesspool tanks and called for typhoid shots. Red Cross workers brought more supplies on Saturday. The Red Cross financed construction of new homes for the poor. Natividad Venegas got a tiny new house, courtesy of the Red Cross, on the same spot where her home had been washed away.

But not everything the Red Cross did met with approval.

Workers tried to determine who would need aid. They asked Frances and Thomas Corbett how many children were in the household. Thomas was Anglo, and Frances was Hispanic. Mixed marriages were uncommon, and the Red Cross worker said she wanted to see the Corbett children to see what they looked like. Thomas Corbett angrily replied: "They look like children!"[27]

Meanwhile, the search continued for bodies and survivors. By Monday afternoon, sixteen bodies had been located. Eight more were later discovered. Two were never located. Cleanup crews also worked. One man, Russell Lee Thompson of San Antonio, was injured when a rail uncoiled and hit him in the forehead. He was transported by ambulance to Fort Stockton and recovered.

The Southern Pacific had plenty of work to do, too. Floodwaters had destroyed or damaged one thousand one hundred sixty-five feet of bridges at five locations, and washed out fifteen thousand feed of track in fifteen locations within twenty-one miles

Yard tracks at Sanderson were
virtually demolished.
*Terrell County Historical Commission*

west and six miles east of Sanderson. The yard tracks "were virtually demolished." The railroad sent two hundred and fifty-seven workers to the area, trucked in water from San Antonio, Del Rio, and Marathon, and moved clothing donated to flood victims to Sanderson at no charge. The main line was repaired after only ninety-eight hours.

Johnny Williams, the owner/operator of the Sanderson Wool Commission Company warehouse, told the *San Angelo Standard-Times* in August: "I'm just like a ranch man who gets his place paid for and then loses it through no fault of his own. I've had to start over from scratch."[28]

Williams was building a new warehouse—on higher ground—when he gave the interview.

Four years after the flood, the U.S. Soil Conservation Service presented its report, *Watershed Work Plan for Watershed Protection and Flood Protection*, that stated a need for a series of eleven dams. And maybe a visitor driving through the sleepy town—with a shrinking population of less than eleven hundred these days—in the high desert might think that spending $34 million on flood control in this part of the country is a waste of money.

But those who still live in Sanderson and remember June 11, 1965, think otherwise. As Darrel Seidel of the U.S. Department of Agriculture's Natural Resource Conservation Service told the *San Angelo Standard-Times* in 1998:

"We just wouldn't be here without it."[29]

## Chapter 19

# The Wichita Falls Tornado of 1979

## *"There was nothing there"*

**W**ichita Falls resident Cynthia Bush said, "I've always believed in premonitions, and this one came true."[1]

In October 1978 Bush and her mother dreamed of a destructive tornado. "It seemed so real!" said Bush, who woke up in a cold sweat and sat up in bed. Six months later, on what came to be known as "Terrible Tuesday," Bush would relive her nightmare.

April 10, 1979—five days before Easter—started off as a typical spring day in North Texas: hot, humid, barometer low, and a forecast calling for thunderstorms. By the afternoon, several thunderstorms, about ten to fifteen miles in diameter, had spawned in Texas and Oklahoma. Sometime around 3 P.M. a "sandy colored" tornado formed in one of those storms and struck Vernon in Wilbarger County, Texas. The twister struck south Vernon, moving southwest to northeast, ripping through a trailer park, destroying farm equipment, and blowing eighteen-wheelers off U.S. Highway 287. The storm left eleven people dead, more than sixty injured, and forced the town to set up emergency facilities at Central Fire Station. A temporary morgue was set up at the National Guard Armory. Emergency rescue equipment and personnel were sent from nearby Wichita Falls. Within hours, Wichita Falls would be requesting that the equipment and personnel return.

The system moved into Oklahoma, where another tornado struck Lawton, damaging a seven-square-block area, killing four, and injuring more than fifty. Ratliff City and Rush Springs also reported damage in Oklahoma from the storms, as did Harrold,

Lockett, and Seymour in Texas. Another thunderstorm formed southwest of Wichita Falls and began moving toward the city.

Tornado watches began in Wichita Falls around 2 P.M. Around 4 P.M. Cynthia Bush asked her boss if she could go home early and get her children. "I told him I felt like all hell was about to break loose," she said. "Him and all the employees laughed." Bush went home, though, and noticed a green tint to rapidly moving dark clouds as she drove to pick up her fourteen-month-old daughter and three-year-old son at the babysitter. When she left with her children, the babysitter was pulling a mattress into the hall for her and her own children to crawl under.

Instead of going to her home on McNeil Avenue, Bush drove to her parents' house, also on McNeil but closer to the main thoroughfare Southwestern Parkway and on a solid foundation. When she arrived, her parents were watching the news for updates on the storms. At 6 P.M. broadcasters announced a tornado warning for Wichita Falls. At that time, warning sirens went off, and the television station reported that a tornado was on the ground at Memorial Stadium and residents should take cover. Bush and her mother walked onto the porch, but Bush couldn't see a funnel. "Look at the white clouds on each side," her mother said. "That black is the tornado and it's moving closer."

Bush and her mother ran inside and began opening windows. Bush dragged a mattress into the hall, ignoring her father's chides of being a "scaredy-cat." She took off her glasses and her children's glasses and placed them on the coffee table, then slid under the mattress with her children and mother. Her father was on the porch, watching, as the noise grew closer. Suddenly, Bush's father shouted, "OH MY GOD!," slammed the door, and took shelter under the mattress with his wife, daughter, and grandchildren.

Bush heard the roof being sucked off the house, windows and the patio door exploding. Then everything became still. Recalled Bush:

> My father said "well it's over" and right after I told him it wasn't and to get back down. The other side of the storm hit us. It's strange but the only prayer I could think of was the 23 Psalms. As soon as the storm

passed, jagged hail fell that looked like sharp pieces of glass. Then nothing.[2]

Bush crawled from under the mattress. The house had been demolished, but she and her children were all right, "just scared to death." Her mother had a broken collarbone, apparently from a chair that flew down the hall, and her father had a deep cut on his arm that had held down the mattress, but at least they were alive. Bush and her parents craved cigarettes, but her mother smelled gas. "That was probably what kept us from blowing everything up," Bush said. She found her glasses, and her children's—all undamaged—on the coffee table.

> We got ourselves and all the neighbors we could find, put them in the next door neighbors' car (after we unburied it), and picked our way around the debris and electrical wires to my house. Which by the way didn't get hit at all.

Looking down McNeil Avenue after a
destructive tornado hit Wichita Falls in 1979.
*Moffett Library, Midwestern State University*

The house of Ann and Troy Bryant wasn't so lucky.

Troy, a sportswriter, was working at the *Wichita Falls Record News* that evening, and Ann had worked late at the bank. After work, she signed up her six-and-a-half-year-old daughter, Jan, for swimming lessons at the YWCA. When she got home, she planned to go to the grocery store, but Troy called from the newspaper and told her to wait. He said there had been a bad storm up in Vernon and rain and hail were coming their way. After hanging up, Ann went outside and pulled their car up the driveway for better protection. There was a six-foot privacy fence on the left, a storage shed in front, and the side of the house on the right. Afterward, Ann went back inside, throwing a few items into a laundry basket in case she and Jan had to go to the storm cellar: can of tuna (but no can opener), box of crackers, jug filled with water, cups, and Jan's pillow. She had the stereo on, Jan sat watching television, and Ann began ironing a pair of jeans. At about 6 P.M. the phone rang. It was Troy.[3]

"Go to the cellar," he said.

"Why?"

"Don't argue. Just go to the cellar."

"Why?"

"Go to the cellar now," he snapped and hung up the phone.

Twenty years later, Ann still laughs at the conversation. "In twenty-seven years of marriage, the man has never hung up on me except that one time."

Ann pulled on the jeans—one leg ironed, one not—grabbed the laundry basket, and took Jan outside, around the house, and down into the storm cellar. Once downstairs, however, she realized she didn't have a flashlight or shoes. After telling Ann to stay in the cellar, she went back into the house, found a flashlight and pair of shoes, and returned to the cellar. "And I realize I don't have my purse. You know us women, we have to have our purses. So I again told Jan, 'Be calm,' and I make a mad dash back in." Finally, with purse, shoes, flashlight, and daughter, Ann shut the door to the cellar.

Their cocker spaniel puppy, Monte, began whimpering at the door. Ann was allergic to dogs and didn't want to be with a dog in the confined space of the shelter. She began singing songs with her daughter—*Farmer in the Dell, Old McDonald, Jesus Loves Me,* the ABCs song—trying to keep Jan occupied and not frightened. Jan, however, didn't like being in the shelter, wanted to know why they were there, why her mother wouldn't let Monte inside. . . . During it all, through what seemed like hours, Ann Bryant never heard the warning sirens.

The warning system worked. Wichita Falls had twenty-eight warning sirens around the city and they had been tested April 2. The first went off fifty minutes before the tornado. At about 5:50 P.M. Don Hart, chief of fire communications, received word that a tornado has been spotted just west of Memorial Stadium. He sounded a three- to five-minute blast on the sirens, and again at 6 P.M. and 6:08 P.M. as the tornado moved through Wichita Falls.[4]

Meanwhile, in the storm cellar at the Bryants' home:

> There was a little six-inch by nine-inch window. I swear that the sky cleared off. It had been an overcast day, yucky looking. But the sun came out bright and I heard birds singing. They told me I'm crazy, but I swear this is what I experienced.

If the sky cleared, it didn't stay that way for long.

Their house sat about two blocks from Southwest Parkway. Ann heard a rumbling—"I thought a bunch of semis were coming down Southwest Parkway," she said—and the sky darkened. Monte started scratching at the door, and Jan began crying to let the dog in. Ann opened the storm door, but the dog evaded her. "I can't catch her without leaving the cellar and I don't want to leave Jan." It had started to rain. The roar got louder. Ann left Monte outside and went down the stairs.

> I didn't know that the rope hanging down from door was so I could tie it to the hook down in the bottom in the cement. So the door raises up to ninety-degree angle, and it closes. Jan and I are sitting in lawn chairs just straight across from the door at the base of the stairs. So I know enough when the door started whipping around to move into the corner opposite the stairwell but on the same wall. I got Jan down in the corner, put a pillow over her head—all

these things you're taught to do in the hallway at school.

The wooden door opened to a forty-five-degree angle and slammed shut again. This time, the door splintered lengthwise. As the roar went over the house, pink fiberglass drifted into the cellar. Ann's first thought was that the storm had caused some roof damage. As the roar went over the house, the sky got a little lighter and Ann heard voices. She went up the stairs and poked her head out of the cellar.

> These men are coming through and they ask us if we're OK and how many of us there are. They're kind of taking an inventory of the neighborhood and everybody. I didn't think anything of getting out of the cellar. Jan and I climbed out and looked around. I was very startled to see, yeah, I had some roof damage. My roof was laying on the ground, halfway covering the door of the cellar that I was trying to get out of.

Troy later told Ann he had no idea how she and Jan crawled over the debris.

Once outside, men from the neighborhood were checking on cars, looking for ways to transport injured people to the hospital. Ann and the men decided that her car, despite the windows and windshield gone, would run.

By then, Ann Bryant realized the storm had been a tornado.

> I guess there wasn't anything to see. There was one closet of my house standing and some walls around a bathroom at a crooked angle, and I could see all the way to Southwest Parkway when there had been two blocks of houses between me and there.

A few houses were left with only minor damage, but most of the neighborhood was gone.

"There was nothing there," Ann said.

> I don't think the destruction ever hit me so much as well, we're OK. I wondered what happened elsewhere in town. Is Troy OK? One car came down the street and said downtown's OK. The next vehicle that

came through would say, "My sister was on the phone with her boyfriend and he was working at such-and-such downtown and he said it was coming right at them." I couldn't get any reliable reports on what had happened outside my immediate area.

Of course, Troy had been at the newspaper on the rooftop watching the storm. Ann decided to stay put, figuring her husband could find her there. Troy, however, had planned to get gas after leaving work. Now that was impossible, so he was waiting for his wife and daughter to come to the office.

At 7:15 P.M. it began to get dark and started to rain some more, so Ann was persuaded by neighbors to let them take her and Jan to the newspaper. Before leaving the area, they met Troy and fellow journalist Doug Brown coming in. Troy stood in the driveway, stared at where the house had been, and cried.

Said Ann: "I think that's when it finally hit me how lucky I was and that we had lost everything."

They returned to the newspaper. Power lines were out all over town, telephone lines were down, water lines were out, gasoline couldn't be pumped. At the newsroom, Jan began to go into minor shock.

They laid her on a desk and covered her with coats. About that time, Jan's mother and father and aunt and uncle walked into the newsroom. Her uncle was wearing a AT&T baseball cap—a gift from his daughter who worked for the phone company—so the journalists thought he worked for the phone company and was there to fix phone lines. They were trying to figure out how to put out a newspaper the next morning and not miss the biggest story in years. It wouldn't happen though. The *Record News* wouldn't be able to publish until Thursday. The *Times*, however, put out an afternoon edition Wednesday that was printed by the *Dallas Times Herald*.

Ann Bryant's parents, aunt, and uncle had arrived in Vernon just minutes after the tornado hit. Her parents, who lived in Quanah, had gone to Vernon to sign papers to buy a new combine. The combine they planned to buy, however, was upside down in the Highway 287 median. They quickly called one of Ann's cousins in Wichita Falls. She told them: "It's going to miss us, but if it doesn't get Ann it's going to be a miracle." They piled back into the car and drove to Wichita Falls.

The National Guard, quickly ordered in by Governor Bill Clements, was on the scene by the time they arrived, preventing them from getting to the Bryants' house, so they drove to the *Record News*. It was decided that Ann and Jan would go to Quanah and return the next morning with a pickup and stock trailer to see if there was anything to salvage at the house. Somehow, they made it back to their house before leaving town. They picked up Jan's bicycle and found the dog, Monte. A neighbor, two or three blocks away, found the cocker spaniel puppy wandering around the mangled neighborhood. Ann, Jan, and her relatives left Wichita Falls late that night. They stopped at Vernon to eat. Ann said:

> I went to the ladies room, which was apparently right next to an air-conditioning unit, and the air-conditioner kicked in and I about came unglued. I think it was the only time I got scared or cried. There was something about the roar of it that scared me.

Two other lucky survivors were Petty and Chet Sutherland, who lived in an apartment complex near Midwestern. They had watched the TV weather bulletins, determined the storm was moving their way, and decided they had better leave. When they got outside, the storm was very close. They jumped in their car and fled, with the tornado right behind them. Trying to outrun the storm, the two knew they were losing ground, but they managed to pull over and stop at an underpass as the tornado whipped by. When the Sutherlands returned to their apartment, they found the front of the building ripped off and an RV overturned. They figured they probably would have been sucked out of the apartment had they stayed.[5]

The twister lasted thirty minutes and left an eight-mile swath of destruction about nine blocks wide. The width of the tornado ranged from one-quarter to three-quarters of a mile. It sucked the press box from the walls at Memorial Stadium, churned along Highway 79, and trapped shoppers inside the Sikes and Southmoor shopping centers. Cars were flattened. Roofs were sheared off houses. Medical supplies were hurried to the city from Fort Worth and Arlington. That night, no one knew how many were dead.

"It wouldn't surprise me if we had a hundred dead at the final death count," Mayor Kenneth Hill said at a news conference.[6]

The entire population of one hundred eleven thousand had no light, drinking water, or sanitation. Of the homes hit by the tornado, eighty-five percent had been destroyed. Only one radio station, KTRN, was left broadcasting, and only because it had auxiliary power. The station kept up a plea for blood from the Red Cross and gasoline for emergency vehicles.

The storm system continued to move. By 1 A.M. new storms formed in a line running from Erath County to Montague County. Tornadoes were reported in Coleman, Wise, and Montague Counties, but there were no reports of injuries or damage. Baseball-size hail was reported in Denton County around 1:30 A.M. Another tornado was reported at 2:06 A.M. in Sherman, and hail fell sporadically between 2:30 and 4 A.M. in Fort Worth. Tornado watches extended to the Dallas-Fort Worth area, but there were no confirmed funnel sightings, and the watches expired at 7 A.M. By Wednesday, the National Weather Service was predicting "storm-free skies and mild temperatures" over Dallas-Fort Worth.

On Wednesday, rescue operations continued in Wichita Falls and people began sifting through their homes.

Ann Bryant returned to her wrecked home that day. Troy and his parents, who lived in Paducah, were already there. Ann recalled:

> It dawned on me I didn't have my wedding ring. I'm notorious for taking off my ring when I'm washing my hands and leaving it somewhere. I had left it on the vanity sometime that afternoon. And my father-in-law got down on his hands and knees with an old piece of window screen and sifted through all the sheetrock debris and stuff in the general area until he found my wedding ring.

There really wasn't much to salvage, however. A three-drawer chest of drawers had been beside a portable sewing machine in the southeast corner of the house. The chest was gone, but the sewing machine was there. A piece of fabric near the sewing machine was found hanging in the trunk of a split tree in a neighbor's yard. Troy's closet on the same wall had a magazine collection still intact. A cabinet over the refrigerator lay on the ground. Inside, Ann found her china place-setting for twelve—with nothing broken. All of the dishes in the dishwasher were OK. "It was so weird," Ann said.

Even weirder?

About a week later, Ann got a phone call at First Wichita National Bank. The woman asked if she had a daughter named Jan Elizabeth born Nov. 26, 1972. Ann said she did. The woman, who lived on the other side of town, said she had found Jan's birth certificate in her yard.

The *Record News* came out Thursday morning, with a banner headline: "Giant tornado lays waste to city." That day, President Jimmy Carter declared Wichita Falls a national disaster area, and three federal disaster assistance administration "one-stop" centers opened Saturday in Wichita Falls and Vernon.[7]

In the end, the Texas storms had left fifty-four dead, one thousand eight hundred and seven injured, and caused $42 million in damages. Forty-two had been killed in Wichita Falls, three thousand homes were destroyed, another six hundred heavily damaged. Most of those killed had been in cars trying to outrun the storm. Fifteen thousand people had been left homeless.

On Thursday, the *Wichita Falls Times* printed this front-page editorial:

Wichita Falls lives.

Wichita Falls has endured a catastrophe that can only be compared to wartime devastation. With prompt action of our elected officials and the famous Wichita Falls spirit of cooperation, Wichita Falls will recover, although the process will require months and years.

The true mettle of our citizenry is being put to the test. With about 10 percent of the homes in Wichita Falls demolished or severely damaged, there is hardly a person in this city without a relative or friend who suffered great loss. It is no hyperbole to say that this adversity has affected every one of us.

Perhaps once in a lifetime, comes a time to show our neighbors and the world what kind of stuff we are made of. This is that time. This is the time to help those who have suffered loss, and a time to cooperate with declarations of emergency procedure.

Evidence of such positive spirit was apparent within hours after the disaster. With the tremendous cooperation of the *Dallas Times Herald*, the *Times* was able to publish Wednesday afternoon an edition packed with information concerning the storm and the resulting cleanup.

At a time of crisis, the list of heroes is long, too long to be enumerated here. But the gratitude of the community should go to radio station KTRN, the only broadcast medium with auxiliary power which enabled it to broadcast continuously through the hours of the disaster.

Our neighbors from hundreds of miles are offering their help in our time of trial. Of that, we are truly appreciative. Match that spirit of neighbor helping neighbor and Wichita Falls will, in time, overcome this natural disaster just as it has overcome lesser disasters in past years.

Yes, Wichita Falls lives.[8]

But there were some setbacks.

Such as charges of price gouging. Lumber that sold for forty cents a board foot a week before the storm was reportedly selling for eighty-six cents a board foot a week after the tornado. Prices for gasoline and food were also reportedly hiked. The city council passed an ordinance Wednesday setting a fifteen-day price ceiling on rent, gasoline, groceries, and other items. A few people were arrested for suspected looting.

But for the most part, the stories were positive. Gas station owner Steve Brewer, who lost his house and a rental house, sold gasoline at his service station for ten cents less than the normal price. Public schools, except for the damaged Rider High, McNeil Junior High, and Cunningham and Ben Milam Elementary, opened Monday. A district judge mounted an American flag on what was left of his home.

Ann and Troy Bryant had owned their house for nine months. Now they had to rebuild. The editor of the *Record News* let the Bryants live in an extra bedroom for a few weeks as they tried to return to a normal routine. Meanwhile, the government agreed to bring in mobile homes for tornado victims. A park was converted into a mobile home park. Victims could live there or put the mobile home on the property where they were rebuilding. The Bryants opted for the latter. Construction started in October. They were supposed to be in their new home by Christmas, but they didn't move in until around Valentine's Day. The insurance declared their car totaled, and they picked out a new Pontiac.

Ann Bryant recalled:

> The insurance company was very good to deal with. But less than a year later when it was time to renew, they didn't renew our policy, which I always thought was a little questionable. I had no control over that claim. If I had had two-three wrecks and they canceled me that would have been one thing, but to cancel me for a claim that was out of my control I thought a little. . .

But overall, Ann said, the rebuilding and regrouping process wasn't a nightmare. The Red Cross gave out vouchers for one suit of clothes apiece. They provided pots and pans, dish linens, one set of sheets, and a blanket. "Things were good," Ann said. "People rallied."

For Ann Bryant, the worst came on the first anniversary of the storm. The media were doing their one-year-later stories, how everybody had bounced back, how great life was in Wichita Falls. Husband Troy, a media junkie, watched the TV documentaries, read the articles, and couldn't understand why Jan and Ann didn't want to relive April 10, 1979.

The anniversary came on another dark day. Ann was working at the bank as lead teller. At about 1 P.M. the weather sirens went off. The employees went to the basement for fifteen or twenty minutes, then decided it had been a false alarm. They went back upstairs, got everything ready, and the sirens sounded again. Again, they had to secure the premises and go back into the basement. "I'm wanting to go get my kid so we'd be together and they're saying, no no no, you can't leave the building," Ann said.

It turned out to be another false alarm. Back upstairs. Back to work. But the warning sirens cried out for a third time. Ann recalled:

> This is all in a matter of a couple hours, so they decide to close the bank. And the big boss makes some joke about you don't want to be in Ann's way when those sirens go off. I think that was harder on me than the day of the actual tornado. Just the possibility of it. Plus, I had had a nightmare the night before about the tornado.

Like most cities, Wichita Falls recovered. Still, the tornado had a lasting effect. Ann Bryant said her daughter "was terrified of storms for a long time. She still has trouble with storms. They still really scare her."

And Cynthia Bush hasn't forgotten the 1979 tornado—or her nightmare six months before the storm. She wrote:

> It's been twenty years since Terrible Tuesday and we're overdue for another one. I keep watching the weather, the sky, and listen to my premonitions.[9]

## Chapter 20

# The Saragosa Tornado of 1987

### *"I didn't know a tornado could do so much damage"*

In Far West Texas, Saragosa postmaster B.R. Begay points out, "sometimes severe thunderstorms erupt so abruptly that one has little or no time to react."[1]

I've driven down state Highway 17 dozens of times on my way to Fort Davis or other points in the Big Bend country. Usually, I'm lost in thought, plotting out a Western novel to be set in the Davis Mountains, but whenever I drive past the small village near Toyah Creek about thirty miles south of Pecos, I'm brought back to reality. I remember 1987.

Once, while on assignment for *Boys' Life* magazine, I stopped at the Saragosa Cemetery. The date on many tombstones puts everything in perspective.

May 22, 1987.

Beginning as a farming community in the 1880s, Saragosa had never boomed. In 1987 its population stood at just more than one hundred eighty. At 8:15 P.M. May 22, a multiple-vortex tornado caught residents by surprise. No one had expected a tornado. Only twenty tornadoes had been spotted in sparsely populated Reeves County between 1950 and 1986—causing no fatalities.[2]

A severe storm twenty miles northwest of Balmorhea had been picked up on radar earlier that afternoon. Hail was reported shortly before 4 P.M. By 7:20 P.M. the storm, now twenty miles southwest of Pecos, had intensified, and severe thunderstorm and flash flood warnings were issued. Thirty-four minutes later a tornado warning was put out for Reeves County. An officer with the

Pecos Department of Public Safety spotted a tornado four miles west of Balmorhea at 8:05 P.M., and a tornado was reported at Saragosa at 8:14 P.M. The twister had formed a mile southwest of the community and moved northeast, a half-mile to a mile wide. The twister would be given an F-4 rating on the Fujita Tornado Scale, packing winds between two hundred seven and two hundred sixty miles per hour.

At Saragosa Hall, a graduation ceremony had begun at 7 P.M. for preschool children. An estimated eighty to one hundred people, including twenty children on stage and another thirty or forty children in the audience, were in the hall when the twister struck. The steel-reinforced concrete south wall collapsed. Parents tried to protect their children. Many hid under tables holding refreshments.

Joey Herrera, a Pecos school board member, had come to speak at the graduation ceremony. He brought his wife, Elsa, and one-year-old son Jonathan with him, although he had told his wife she didn't have to come. When he heard the panic cries of "Tornado!" Herrera stopped his speech and stepped outside to see the twister moving toward the hall. He ran back inside, tore off tablecloths to cover Elsa and Jonathan, and huddled against a wall. The building shook, and that wall collapsed. Forty minutes later, Herrera was pulled from the rubble and taken to Reeves County Memorial Hospital. His wife and son had been crushed to death.

Of the thirty people killed by the tornado, twenty-two died at Saragosa Hall, including two children only one year old and two others less than a year old. Elsewhere, four people died in mobile homes, three in frame houses, and a seventeen-year-old girl was killed in a vehicle. The other victims' ages ranged from two to seventy-seven. All but two of the thirty killed were Hispanic.

A fifty-three-year-old Saragosa man described the storm for the *Dallas Times Herald*:

> It got very dark, like night. I got my family into my niece's Blazer, to try to beat the storm. Gracie was driving, and I was sitting on the right side, and my wife was sitting between us, holding little Mario. We were going to pick up my mother-in-law ... and then get away from the storm.

A memorial at the Saragosa Multipurpose Center lists the
names of the thirty people killed in the May 22, 1987, tornado.
*Author's Photo*

[After being caught by the tornado], I thought we
had better leave the car and get to some shelter. But
when I opened the door, the thing came in and just
ripped it wide open, and it was trying to lift the Blazer
off the ground. Just as we were about to go up into the
air, the tornado slammed another car into my side of
the Blazer, and another into the driver's side. We were
caught in the middle, like the meat in the sandwich.
So the tornado couldn't pick us up. The three cars
together were too heavy.

We got as low as we could. And it's a good thing
we did, because a big piece of tin from somebody's
roof crashed through the windshield. If we had been
sitting up, it would have cut our heads off.

It was very dark. We couldn't see anything. . . . We
couldn't hear nothing but the screaming of the storm
and the hail on the roof. I don't know how long we

were there. It must have been just a few minutes, but to me it seemed like three or four passed, we couldn't get out of the Blazer because we were covered with boards and dirt. Then our neighbor . . . got some people to help him, and they came and got us out.[3]

More than one hundred and sixty people had been injured, and eighty-five percent of the buildings in the village were destroyed. Another survivor told the *Dallas Times Herald*:

I didn't know a tornado could do so much damage. I saw the community center fallen down, and the people in it. I saw a man lying over there across the road. I saw somebody's arm, torn off, lying on the ground. I and my sisters were crying, because we thought our grandfather and grandmother were still in the house.[4]

Rescue efforts began immediately. Many people remained trapped at the hall. By 10:30 P.M. Reeves County Memorial Hospital in Pecos could hold no more patients, so other victims were transported to Monahans and Fort Stockton. Funeral homes were swamped. Pecos Police Chief Ed Krevit told the *Dallas Morning News* that Saragosa had been "utterly destroyed."[5]

That weekend, volunteers arrived from Odessa, Lubbock, Pecos, and elsewhere across the state. More than forty American Red Cross workers arrived, and priests came from more than fifty nearby parishes to help counsel victims and oversee funerals. The Dallas Salvation Army sent volunteers and a truckload of clothing.

The Permian Basin Humane Society in Odessa searched the debris to rescue pets and livestock. Goats, dogs, ducks, chickens, and other animals were found alive. Scott Armstrong, the animal rescuer, also found the carcasses of several pigs, horses, and dogs. But cats, dead or alive, were hard to find. "I don't know where the cats went," Armstrong told the *Dallas Times Herald*. "Everybody had cats. Maybe they're lighter and more agile."[6]

Damage estimates ranged from a low of $1.3 million to a high of $8.7 million. The *Dallas Times Herald* put damage estimates between $3 million and $5 million.

But Saragosa didn't give up, and it didn't die.

A grim reminder of how deadly tornadoes can
be: tombstones at the Saragosa Cemetery.
*Author's Photo*

One resident raised an American flag over a wrecked adobe
building. "It is to give some hope," he said.[7]

On Tuesday, Governor Bill Clements visited Saragosa, but he
took criticism for waiting until four days after the storm and for not
sending out the National Guard to prevent looting Friday night.
The same day, President Ronald Reagan declared the village a
disaster area.

At a funeral Mass on Tuesday, El Paso Bishop Raymundo Pena looked across Saragosa Cemetery with its monuments and wooden crosses. Seventeen caskets sat on plastic milk crates.

Pena said:

> Tomorrow we begin to rebuild our homes. We will overcome the difficulties of this moment. We will stand up and we will walk again.[8]

Chapter 21

# The Fort Worth Thunderstorm of 1995

## "Thousands of light bulbs were falling from the sky"

**S**o, you say hail can't reach the size of a softball?

I beg to differ. I can laugh about it, too.

Maybe that's because Texas journalists always seem to find humor in weather.

In May 1884 the *Presidio County News* noted that a recent West Texas hailstorm "had no respect for even the temple of justice, and its roof is full of holes" like a pepper box.[1]

And when severe thunderstorms pounded the Panhandle in 1886, the *Tascosa Pioneer* reported: "It is not understood just what brought about these storms ... unless it was a job put up by the glass men and the repairers, or else the storms came along to get a notice in the newspaper."[2]

People are often amazed at the condition of my Nissan Pathfinder and its two hundred thousand-plus miles. "That's a 1991 model?" they say. "It certainly doesn't look that old."

I smile and reply, "Four thousand dollars in repair work after a hailstorm can do wonders."

May 5, 1995 had started off as another evening of work at the *Fort Worth Star-Telegram*. The Mayfest celebration downtown was going on. Parking was hard to find. I read sports copy, cussed, probably slammed a few phones, and occasionally glanced at the television. Springtime in Texas meant severe thunderstorm warnings, and this was no exception. The radar map was red.

The sports department at the *Star-Telegram* sat directly above downtown Sixth Street—literally. The sports editor's office was a pretty good place to watch fender benders. But it wasn't a good place to be during tornado warnings. It began to hail. You could hear it pounding on the roof and walls. At some point, the newsroom was evacuated to the basement. I ducked out the Sixth Street door—where it was covered, mind you—with a few other colleagues and watched it hail.

This wasn't ordinary hail. Some had a diameter of 4.5 inches. A co-worker's wife called the office to say she had picked up a hailstone and put it in the freezer. It was huge, like a softball, with spikes like a mace.

Downtown, softball-size hail pounded the sidewalk like sledgehammers. I didn't think about the ten thousand people outside at Mayfest. I thought about my Pathfinder. Then I went into the basement. I should have tried to move the sports utility vehicle into a covered parking area. Now it was too late. Colleague Tommy Cummings, however, didn't think so. He went out to save his Chevrolet Cavalier.

"My first recollection was how it seemed like thousands of light bulbs were falling from the sky and exploding on the pavement," he recalled. "It wasn't like the hail I knew in Oklahoma where it was like a bunch of marbles bouncing around. These babies were throwing off shrapnel when they hit pay dirt."[3]

Another journalist opened his *Star-Telegram* umbrella and stepped outside. A giant stone ripped through the umbrella. He went back inside, put away the umbrella, and gave up. Cummings, on the other hand, thought about "my insurance deductible and the current valuation of my car. It didn't add up very well, so I decided to risk it and rush to my car, which was about fifty yards away in a church parking lot east of the *Star-Telegram* building."

Cummings covered himself with a box on his shoulders and ran into the melee. A car windshield in the parking lot caved in. "After another few yards, I heard glass crashing. I peered from beneath the box. It was the Bank One building to the north."

The cardboard box, now wet, didn't offer much protection. A hailstone ripped through it. Shards of hail bouncing from the pavement stung Cummings' legs. Finally, Cummings reached the Cavalier and grabbed his keys as the wind swept away his soggy box. Another stone knocked a piece of trim from the car.

Cummings climbed inside, saw that his windshield was cracked, and pulled the Chevy into the garage.

"The *Star-Telegram* parking lot was covered with windshield glass," Cummings said. "The pavement had tiny pockmarks from the stones. . . . The good news is that I stayed up all night, got to the insurance adjuster when he opened his doors and got a check for my car way before anyone else did. The only problem was that he didn't buy my story that the hail had knocked the rubber off my worn tires."

They finally let us out of the basement so we could survey the damage. My Pathfinder looked like a NASCAR Winston Cup car after a bad night at Bristol. The windshield had been smashed. Body damage was extensive. Heck, it looked as if someone had taken a ball-peen hammer to every square inch of the body. But it was drivable. Insurance adjusters would declare many cars totaled. Another colleague asked me to get her umbrella out of her smashed pickup truck. When I sat at my desk, I noticed my bleeding hands.

Results of the 1995 hailstorm in Fort Worth: about four
grand worth of body damage to this 1991 Nissan Pathfinder.
*Author's Photo*

I didn't, however, put in for workman's comp.

I was lucky. Well, not that lucky. About six weeks earlier, I had backed the Pathfinder out of my garage in Dallas to load up for a vacation out to the Big Bend. Just in time for it to hail and leave the roof and hood full of small dents. An insurance adjuster wrote me a check—less my $500 deductible. But I hadn't started the body repair when the May 5 storm doubled, tripled, quadrupled . . . the size of my March hail dings. That meant I was out another $500 deductible.

A hailstone had punched a hole in the hood of one car in the parking lot.

Then there was the case of Steve and Valerie Kaye. They lost both of their cars in the hailstorm. Said Kaye:

"Total repairs were more than $16,000. Now that's a great memory. Valerie was eight months pregnant at the time. That made it extra nice. We wondered how we would get to the hospital if the baby decided to come."[4]

It took months for hail victims to get the body repair work complete. Body shops, glass repair, and roofing companies stayed tremendously busy.

Rain, hail, and seventy-mile-per-hour winds took a destructive and deadly toll on Dallas-Fort Worth. Twenty people died across Texas—fifteen of those in Dallas County, most of them drowning victims, when the storm moved east and the hail turned into torrential rain. At least sixty people were injured at the Mayfest event in Fort Worth. The National Guard armory became an emergency center for broken bones (by hailstones), cuts, and other injuries. Winds blew the roof off the Lillian Norris girls dormitory. Fort Worth suffered $16.8 million in damages to city buildings, vehicles, and equipment.

Damage to the Will Rogers Equestrian Center hit $1.38 million. City Hall suffered $493,000 in damages, and the Amon G. Carter Jr. Exhibits Hall had a $515,000 tab.[5]

The storms caused an estimated $1 billion to $2 billion worth of damage in Texas and New Mexico, the bulk of that in the DFW area. A weather program on The Discovery Channel called the Fort Worth storm the costliest thunderstorm in U.S. history.

It also ranks as one of the ten costliest natural disasters in U.S. history.

Maybe that's why, for the rest of my tenure at the *Star-Telegram*, whenever the TV weather maps showed red and orange, and the meteorologists said, "It looks like hail," I didn't wait to move my Pathfinder to covered shelter.

And I had a lot of company.

## Chapter 22

# The Drought and Heat Wave of 1998

## *"The country was hurt"*

**S**o, how hot was it?

Grapevine resident Dan Langendorf answers:

> My yard was reduced to burnt grass. We never went outside unless we had to. I started wearing shorts and T-shirts to work, a habit I now find hard to break. I couldn't teach [daughter] Grace to ride a bike because neither one of us wanted to be outside in the evenings. Popsicles melted off the stick before you could eat them. It was impossible to barbecue outside because you'd be drenched in sweat just to flip a couple of burgers.
>
> Our dog went outside to do her business, came back in the rest of the day, a habit she now finds hard to break. We went swimming a lot, but the water didn't cool you off. It was warm, tepid, and I continued to sweat while in the water. . . . The malls were packed because people could go inside and cool off. In fact, there were so many health alerts for the elderly and lower income folks that it was suggested they spend time at the malls during the day in order to stay cool. Older people and lower income folks died because they had no air conditioning.
>
> Speaking of air conditioning, our bills were over two hundred bucks a month—and that was pretty good. I know some people who were paying five

hundred and more to keep cool. I couldn't walk across the street barefooted without literally burning the bottoms of my feet. All our plants died, except for the tree in the front yard, and it looked thirsty all the time. There was no point to watering your yard. Why bother?

People's tempers were always short. They walked around in hot, dazed states, perpetually sweating, cranky, tired, carrying water bottles with them.[1]

Tarrant County went through twenty-nine consecutive days with temperatures one hundred degrees or hotter—and fifty-six days total. Dallas sweated through its hottest summer since 1980.[2]

By the time the heat wave and drought broke—with severe flooding—one hundred thirty-one people had died because of the heat across the state, agriculture losses were estimated at $2.1 billion, and President Bill Clinton had declared Texas a disaster area. The average temperature for the year in Fort Worth was 69.6 in 1998—more than four degrees above normal. Even the low temperatures were above normal.

The mercury rose to one hundred degrees in Dallas on May 6, and that was only the beginning. At 4 P.M. July 12 the temperature hit one hundred and ten—breaking the July 12 record of one hundred and eight set in 1954. On August 4 it rained—barely—in Dallas, and the temperature reached only ninety-four degrees. The streak of one-hundred-degree (or hotter) days ended at twenty-nine.

But summer continued. And so did the heat wave.

September 1998 became the warmest September in the world, with a mean global temperature of 59.98 degrees. It was also the warmest in the United States, with a national average temperature of 59.1 degrees. And in Texas? The mercury rose to one hundred and five degrees in Dallas on September 3—the normal high for the month is eighty-eight.

Nothing was immune. Not even baseball.

Kurt Iverson, official scorekeeper of the Texas Rangers, watched umpire Durwood Merrill use modern technology to keep cool behind home plate. "He wore a silver collar that fit on the back of his neck and kept a steady breeze of cool air circulating through a fan that was built into the collar," Iverson said. "He swore that it worked like a charm, but he looked like he might be the robotic umpire, late for his Halloween costume party."[3]

The Ballpark in Arlington suffered, too. There was a brown patch of dead grass in right field. In 1999 new strains of grass were planted.

Farms, forests, ranches took a major hit.

The drought cut East Texas forests by almost $342 million, a loss that would impact the state economy by an estimated $1.1 billion. Sixty-five percent of pine seedlings planted in 1998 died.

Cotton producers suffered $704 million in direct losses with a harvest forty percent less than the 1997 crop and twenty-nine percent less than the drought-marred 1996 crop. Corn losses reached $255 million. Sorghum losses were at $140 million.

Only winter wheat escaped the drought.

By early August, livestock producers had suffered $451 million in losses. Hay prices went up. Forty- to fifty-pound bales rose from $4 or $5 to $7. Large round bales that usually cost $75 to $85 per ton skyrocketed to $125 to $185 a ton.

Some ranchers were fortunate. Some weren't.

Jay O'Brien, managing partner of the historic JA Ranch, recalled:

> The JA felt the heat of 1998 more than the drought on the part of the ranch I run. It was blessed with a couple of freak rains that the rest of the area missed and weathered 1998 better than the rest of the Panhandle. The drought of 1996 hit us worse. We went from May to May without a rain.
>
> Jiggs Mann was not as lucky. He had a tough fire that lasted for days. I drove through his lease from the south, amazed at how dry it was. As I crossed from his pasture where the fire was burning into our Churchill pasture, I crossed from drought to green grass. Rain acts that way.
>
> I have always felt that fire does a lot of good; however, you need rain after it. The cedars were set back

by the fire, but Jiggs did not get the follow-up rains, and the country was hurt.

The other two ranches I run, the Exell north of the Canadian River in Potter County and my ranch between Clarendon and McLean, did not fare as well in 1998. Being primarily yearling operations, we were able to ship and not abuse the land too much. The heat hurt the performance of cattle on all of the ranches and badly damaged breeding percentages. May was the hottest ever with half of it recording temperatures over 100 and several days over 110.[4]

In San Angelo and elsewhere in West Texas, older residents remembered the brutal drought of the 1950s. And the young ones? Said novelist Elmer Kelton: "Some younger people who don't remember that drought got at least a refresher course last year."[5]

Farmers had poor crops. Range conditions suffered, and ranchers had to cut back on stock. In some spots of West Texas, the drought continued into 1999. Kelton said:

"A lot of people had to continue to feed through the summer, and a lot are still doing it through this summer. It's still dry out there. It's better this year, but you don't see any rush to restock."

But this is West Texas, where droughts aren't rare. As Elmer Kelton says:

"We're subject to those things over and over again. But each time it comes as a total surprise to a lot of people, or at least it seems to."

# Epilogue

Tornadoes. Drought. Hail. Hurricanes. Flash floods. Raging windstorms. Well, at least Texas doesn't have earthquakes.

Scratch that.

On April 13, 1995, Alpine shook, rattled, and rolled when a 5.6 magnitude earthquake hit West Texas.

Juliette Forchheimer Schwab had spent more than thirteen years in California, but hadn't expected an earthquake to strike far West Texas. At first she thought a train had derailed, then that it was her neighbors arguing. Up north at Big Spring, Doctor John R. Key felt the temblor but shook it off as "a bunch of semis going down the highway." Finally, Juliette realized what had happened and started making phone calls. Her husband, Gregory, thought it was "the coolest thing." She was furious at him.[1]

So Texas has earthquakes, too. Rarely, but earthquakes nonetheless.

But people live with the elements. Like the tired joke about waiting for the weather to change, Texans develop a sense of humor about hail, hurricanes, drought, and tornadoes. Charles Clines tells this story about a twister in the 1960s at Wichita Falls:

> I was playing golf with three other guys (actually there were about twenty of us playing in foursomes and if I'm not mistaken, there might have been a few wagers going) at Weeks Park Golf Course. We saw these gray-blackish clouds forming, which looked like about five miles from us, and then started noticing they started to gain some circular motion in them. We started planning where we'd go if indeed a tornado did form, and the most likely place would have been down in the creek beds. We played about a hole, and *two* tornadoes came down out of the clouds—we were figuring three to five miles from us at the most. They started moving, it seemed, northeast . . . which wasn't toward us. After a few minutes, one of the tornadoes went back up, but the other kept going and eventually

formed one of the classic ones. We had a couple guys in our group worried about the direction it was going because their families lived that direction. No one quit, of course, but the pro came out on a golf cart to take a couple to the pro shop so they could call home to see if everyone was all right. Thankfully, no one's homes were hit. So, we kept playing. We didn't know how much damage had been caused until later that afternoon when we watched the news.[2]

Hey, golf comes first in Texas.

For all of the bad weather, there's sunshine and blue skies. People golf year round. Texans pull off the interstates and highways in early spring to take photos of their loved ones in the bluebonnets blooming on the side of the road.

In 1841 William R. Dewees wrote that although droughts are frequent, if a farmer plants early he can raise a large crop. Dewees went on to describe the climate as "pleasant," the early summer heat as "not oppressive," breezes "refreshing and delightful" and winters mild.[3]

Another early settler, Belle Little, who arrived in Texas in 1872, made this comment about the weather:

Notwithstanding the drouths, frosts, cyclones and insects, the climate of Texas as a whole, since I have lived here cannot be surpassed. When the spring comes with its accompaniment of Texas winds and gentle showers, the wild flowers springing up over the prairie with their riot of color while flinging their fragrance far and near, carry anew nature's age-old message of the Resurrection. Fall brings the frost king, who paints his pictures in all his gorgeous shades on every bush and shrub. In the midst of it all sits the yellow golden-rod, which nods serenely as autumn's flower queen. The winter's chilling blast drives all nature's subjects to seek a long siesta in the cold light of a winter's sun. The wild sumac, the red-bud and the cedar trees which grows in profusion in the rocky sandy soil west of Waco when the snow and frost come make a picture worthy of the greatest artist brush.[4]

Texas weather may be unpredictable, but it certainly isn't unique. Humidity? Try St. Louis, or South Carolina in August. Tornadoes? North Texas may be in Tornado Alley, but twisters strike everywhere in the United States. A rare one even hit Salt Lake City in August 1999. Hurricanes? My sister's home near Timmonsville, South Carolina, was destroyed by Hurricane Hugo—and she lives one hundred miles inland. Floods? Look at the Mississippi or Missouri. And flash floods are common throughout the West. Droughts? As I write this, much of the Southeast has been declared a federal disaster area. Dust storms? I had a flight delayed in Phoenix by dust so thick that at first I thought I was watching an Arizona monsoon. Hard winters? In New Mexico, I've seen it snow in October and May. Hail? I've been hailed on in New Mexico and South Dakota. OK, it wasn't the size of softballs.

The bank thermometer says one hundred degrees as I drive through Waxahachie on my last research trip for this book. It's still a steamy afternoon when I reach a diner in the Oak Cliff section of Dallas. I sit at the counter, sipping iced tea and picking at a chicken-fried steak as a busboy on break talks to a customer about the weather.

"I saw them fry an egg on a sidewalk in New York City on TV last night," the customer says.

"It's hot enough to do that here," the bus boy says. "I'm gonna get me an egg and crack it open outside."

He moves to the kitchen, goes outside, and returns. He doesn't say anything, though. I figure he didn't do it. Why waste an egg? I finish the meal, pay the bill, and walk to my truck. On the sidewalk outside the restaurant, I spot a cracked eggshell. I move closer.

There it is. But the egg white and yolk aren't anywhere near sizzling. In fact, the egg is about to run off the concrete step. I crank up the Pathfinder and back away.

Maybe Texas in July isn't that hot.

Or maybe the busboy shouldn't have cracked open an egg in the shade.

# *Afterword*

Esteemed novelist and friend Loren D. Estleman and I stood in a bar (of all places) debating the ten funniest movies of all time. "The thing about this list is you could ask me a week from now and I might have ten other movies," he said. Which got me to thinking about this book.

I could have easily included other storms, other droughts. As I said earlier, this is not a full-fledged history. Some are obvious choices. The Galveston hurricane of 1900 had to be included, as did the 1902 Goliad tornado, the drought of the 1950s, and the Waco tornado of 1953. Some choices were merely personal. As a Western novelist, the Ben Ficklin flood of 1882 and the Indianola hurricanes have interested me for years.

But what about those I left out?

The drought of 1884-86 is significant, but I already had included the Indianola hurricane of 1886 and the blizzard of 1886. That year must have been one miserable year to be in Texas. Besides, I had a wealth of information on the drought of the 1890s. I could have gone back earlier than 1856 (having lived in Dallas, I was amazed that the Trinity River could freeze over), but complete information on Texas weather is pretty hard to come by before the 1880s. It was tough enough filling a chapter on the spring of 1856.

I was often asked about the F5 tornado that devastated Jarrell in 1997, but that seemed too recent. The last thing I thought Jarrell survivors needed is a reporter coming around asking, "What did you feel?" to put in a book. Truthfully, I had reservations about asking about the Saragosa tornado of 1987 and got one response to queries for personal recollections.

Other hurricanes pounded Texas—the 1867, 1909, 1915, 1932, and 1961 storms come to mind—as well as strong storms before and since, but I thought I had enough hurricanes in this edition. Dust storms were frequent in the 1930s, but "Black Sunday" seemed to tell the best story. There were other tornadoes, too, but again I figured I had a wide enough range.

The summer of 1980 had been deadly, but I opted for the heat wave of 1998 to bring the book up to date. I wanted to include cereal king C.W. Post's rainmaking experiments of 1910-1914 but couldn't track down enough information. Besides, I had the government experiments of the 1890s. The Delta crash in 1985 during a thunderstorm at Dallas-Fort Worth International Airport could have easily subbed out for the 1995 thunderstorm. There are other storms I knew about and thousands that I don't—all of which would make great reading but aren't in this book.

That's because this is only a sampler. The book's limitations are my fault. Maybe somebody will tackle a complete history of Texas weather one of these days.

Put me down for a copy.

*Appendix A*

# Texas Weather Through the Years

(**Note**: Reports of Texas weather can be dated to the 1700s, but these are scattered and incomplete. Therefore this list begins after 1870, with the establishment of the National Weather Service, directed by the U.S. Army Signal Corps, on February 9. Texas had fourteen weather stations in 1870.)

| | |
|---|---|
| 1871 | Volunteers from Galveston rescue the crew of the Virginia Dare, who have lashed themselves over the rigging during a June 9 hurricane. |
| 1872 | Fort Concho, near San Angelo, records no rain in March, "a great contrast with every previous March," post surgeon William M. Notson remarks. |
| 1873 | "Generous rain" in May and June helps end the dry conditions brought about during the winter and early spring in North Texas. |
| 1874 | A September 4-5 hurricane rocks Corpus Christi with "winds and waves contending in mad furry for the mastery in the work of destruction." |
| 1875 | Indianola is hit hard by a September 14-16 hurricane that destroys three-quarters of the city and claims one hundred seventy-six lives. |
| 1876 | It's another wet year in North Texas, with a warm winter and mild summer. |
| 1877 | In July and August, Captain Nicholas Nolan's cavalry expedition into the Staked Plains results in disaster as the troopers succumb to heat and the lack of water and are forced to drink blood from horses. Four troopers, twenty-five horses, and four mules die. |
| 1878 | North Texans experience a mild winter and spring, a wet summer, dry fall, and frigid December with temperatures far below normal. |
| 1879 | A drought, described as the "most serious in 30 years," hits north central Texas. |
| 1880 | Corpus Christi and Brownsville suffer through an August 12 hurricane, leaving "Padre Island covered with wrecks" and five people dead in Brownsville. |
| 1881 | Three people die of sunstroke in August in San Antonio. |

1882   A flood destroys the Tom Green County town of Ben Ficklin.

1883   An October tornado in Tyler blows a train from its tracks.

1884   Heavy rains May 20-21 cause violent flooding on the Trinity and Brazos Rivers.

1885   A meteor lights up the night sky on June 6.

1886   Jesse A. Ziegler reports that "the bayous, creeks and even Galveston Bay were frozen over."

1887   A June 5 sandstorm, with winds at more than 50 miles per hour, limits visibility to less than fifty feet in Abilene.

1888   A hurricane pounds the Texas coast on June 16-17, and another storm makes landfall at Galveston on July 5.

1889   A raging blizzard hits the Texas Panhandle in early November.

1890   Texas has seventy-eight weather stations as Congress transfers meteorological services from the Signal Corps to the Weather Bureau.

1891   The Weather Bureau goes into operation on July 1 as part of the U.S. Department of Agriculture. Rainmaking experiments are conducted in South and West Texas.

1892   East Texas tornadoes on December 6 leave four people dead and eighty injured.

1893   More than twenty people are killed and the town of Cisco is wiped out by a tornado.

1894   A series of March tornadoes cause more than twenty deaths and one hundred injuries in North and East Texas.

1895   Houston's streets turn white with twenty inches of snow while Galveston records fifteen inches and Victoria twelve during a freak Valentine's Day storm.

1896   Four tornadoes strike Denton and Grayson Counties, leaving more than seventy dead and three hundred injured.

1897   Port Arthur is damaged during a September 12 hurricane that kills thirteen.

1898   Tulia records 23 inches of snowfall, including 10.5 inches in December, and the lowest temperature of the year, minus 2 degrees on December 10.

1899   Heavy rain causes flooding on the Brazos River, damaging crops and drowning more than twenty people in Robertson County.

1900   The Great Galveston Hurricane of September 8-9, the worst natural disaster in U.S. history, kills six thousand to eight thousand—maybe even more.

1901    A summer heat wave has temperatures climbing to 100 in Fort Worth on eighteen days.

1902    The Goliad tornado kills more than one hundred and injures approximately two hundred and fifty.

1903    College Station records two 110-degree days on July 24 and August 17.

1904    Tornadoes in March, April, and May kill twelve and injure eighty-six in Denton, Hunt, Robertson, Falls, Limestone, Freestone, Hopkins, Franklin, Titus, Wood, Callahan, Shackelford, Stephens, Mills, Wise, and Archer Counties.

1905    The summer's hot in Waco, with twenty-five 100-degree days.

1906    An April 26 tornado kills seventeen and injures twenty in Bellevue.

1907    Crops in northwest Texas are damaged by a "hot wind" on June 29-30.

1908    Floodwaters damaged three hundred homes along the Trinity River bottoms in Fort Worth after heavy rains April 18-19.

1909    A dust storm, with winds gusting to 52 miles per hour, strikes Fort Worth on March 24.

1910    Heavy rains on September 5-6 send a twenty-five-foot wall of water sweeping out of the Leon River, leaving thirteen people dead.

1911    A January cold front leaves temperatures at 19 degrees in Galveston, and 21 at Brownsville and Corpus Christi.

1912    A late hurricane strikes south of Corpus Christi on October 16, sinking a steamer off Padre Island and killing six crew members.

1913    More than two hundred derricks are blown down by a sandstorm near Humble on February 26-27.

1914    July is blistering as Blanco, Greenville, Bowie, Graham, Henrietta, Seymour, and Brownwood record temperatures of 110 degrees.

1915    A new seawall in Galveston, built after the savage 1900 hurricane, helps lessen the damage during an August 17 hurricane that still kills two hundred seventy-five, many of whom did not heed the Weather Bureau's warnings a day earlier.

1916    A hurricane hits Corpus Christi on August 18, killing twenty and causing an estimated $1.6 million in damages.

1917    Statewide, this is Texas' driest year, with only 14.30 inches of precipitation.

1918    The drought that began in September 1916 ends with October rains.

1919    A September 14 hurricane devastates Corpus Christi. Records claim two hundred eighty-six fatalities, but other estimates cited five hundred to six hundred.

1920    Strong winds in Jefferson County on May 15 blow down one hundred eighty-five oil wells and cause $200,000 in damages.

1921    Floods along the lower Brazos and other rivers September 8-10 cause two hundred fifteen deaths and $19 million in property damage. Taylor records more than twenty-three inches of rain during a twenty-four hour period on September 7.

1922    Thirteen people are killed when two tornadoes strike Austin at the same time on May 4.

1923    May 14 tornadoes result in twenty-three deaths in Howard and Mitchell Counties.

1924    Sixteen inches of snow is recorded at Temple during a February freeze.

1925    The high at Brownsville on December 28 is only 31 degrees.

1926    One thousand derricks are destroyed or damaged, with losses at more than $2 million, during a series of March thunderstorms and tornadoes.

1927    Rocksprings is hit by a violent tornado on April 12, leaving seventy-two dead and more than two hundred injured.

1928    Dalhart brings in the new year with a temperature of minus 7 degrees on New Year's Day.

1929    A cold front on December 17-24 dumps 26 inches of snow at Hillsboro, 24 at Clifton, and a trace as far south as Brownsville.

1930    On May 6 tornadoes plague West, East, and South Texas during a twelve-hour period. Total deaths: seventy-seven.

1931    Temperatures range from a low of 4 degrees at Miami on March 27 to a high of 109 at Fort Stockton on June 19.

1932    Galveston is hit by another hurricane, this one on August 13 that kills forty people and injures two hundred.

1933    Seminole records a temperature of minus 23 degrees, matching the record logged in Tulia in 1899.

1934    Drought conditions that began in 1933 continue as the Dust Bowl kicks up.

1935    "Intense rainfall" June 9-15 causes flooding on the Colorado, Llano, and Pedernales Rivers.

1936    Laredo records a temperature of 98 degrees—on January 17.

1937    An ice storm strikes northeast Texas on January 5-12, resulting in damages of $3 million to $4 million.

1938    Fourteen people are killed and nine injured when a tornado strikes Clyde on June 10.

1939    Drought conditions begin in the late spring, affecting crops and livestock, and continue into the following year.

1940    Austin records 14 consecutive days of freezing temperatures from January 15-28.

1941    Statewide, this is Texas' wettest year with 42.62 inches of precipitation.

1942    The Lower Valley region receives thirteen inches of rainfall in June. Summer rainfall totals usually are 1.5 to 2.5 inches a month. Nationwide, the Weather Bureau begins issuing tornado warnings for the first time.

1943    The temperature in August climbs to 119 degrees in Vernon.

1944    An ice storm ravages East Texas on January 13-14, resulting in an estimated $16 million in timber losses.

1945    Mount Locke logs a temperature of 29 degrees on September 29, a record low for the month.

1946    Thunderstorms and hail strike San Antonio on May 15 and May 29, resulting in more than $6 million in property damage.

1947    An April tornado moves through White Deer, Higgins, and Glazier and eventually logs two hundred twenty-one miles in Texas, Oklahoma, and Kansas.

1948    The Trinity River floods in Dallas after heavy rains on February 25, and tornadoes kill three people the following day after hitting Lewisville and Ranger.

1949    Houston records 22.31 inches of rain in October. A May 15 tornado leaves six dead in Amarillo.

1950    Drought conditions begin and worsen as the year continues. Nationwide, hurricanes and tropical storms are given names.

1951    Temperatures climb to 100 degrees or above on eight consecutive days, August 11-18, in San Antonio.

1952    The drought continues to worsen, affecting most of the state.

1953    A tornado strikes San Angelo and Waco, killing more than one hundred, injuring more than a thousand, and causing $51 million in damage.

1954    Dust storms plague the state, and the yearly rainfall averages 18 inches, the worst since 1917.

1955    Seventeen people, including four students at a Kingsville high school, are killed by lightning from May through August.

1956    In the last full year of the disastrous drought, more dust storms plague most of the state and water rationing is common at several cities.

1957    An April 2 tornado makes its way through Dallas, killing ten, injuring two hundred, and damaging five hundred seventy-four buildings.

1958    Rio Grande City records a temperature of 111 degrees on April 24.

1959    The temperature in Dalhart on January 4 dips to minus 21 degrees.

1960    Flash floods on October 28 claim eleven lives in Austin and result in estimated damages of $2.5 million.

1961    Sustained winds of 145 miles per hour are logged at Matagorda and Port Lavaca on September 11.

1962    The temperature drops to minus 13 degrees at Salt Flat in Hudspeth County on the night of January 11.

1963    Hurricane Cindy causes massive flooding September 16-20, damaging four thousand homes in southeast Texas.

1964    Wichita Falls is hit by a tornado April 3 that kills three, injures more than one hundred, and causes an estimated $15 million in damages.

1965    A giant January 25-26 dust storm affects the state from El Paso to San Antonio to Houston and as far north as Dallas.

1966    Late April rains cause massive flooding in northeast Texas, claiming 33 lives and doing $27 million in damages.

1967    Heavy rains from Hurricane Beulah cause major flooding in September and October, claiming forty-four lives and costing $160 million in damages.

1968    The temperature reaches 118 degrees in Pecos on June 29.

1969    A storm hits the Texas coast on February 13, with tides nine feet over normal in Galveston and wind gusts of almost one hundred miles per hour at High Island and Palacios.

1970    Spring tornadoes cut a deadly swath across Texas, killing twenty-six, injuring two thousand, and causing property damage of more than $200 million.

1971    A March 14-15 dust storm starts in northwest Texas and limits visibility to less than a mile in Dallas.

1972    Fifteen hundred homes are damaged during March flooding in Harris County.

1973    A January 8-11 winter storm causes the loss of 150,000 cattle and costs ranchers $50 million.

1974    Ten people are killed in Austin, and three others die in Central Texas, during flash flooding on November 23.

1975    A tornado kills two and injures forty-two in Lefors, destroying one hundred sixty homes.

1976    Travelers must contend with hail that is piled bumper deep in Rusk County during a severe May thunderstorm.

1977    Midland-Odessa records forty-three days with temperatures reaching at least 100 degrees.

1978    Tropical storm Amelia sprays Texas with heavy rain on August 1-4, resulting in twenty-three deaths and $110 million in damages.

1979    Three tornadoes, in Vernon, Harrold, and Wichita Falls, cause fifty-four deaths, more than eighteen hundred injuries, and $42 million in damages.

1980    Temperatures reach at least 100 degrees on sixty-six days in June, July, and August in Dallas-Fort Worth. Across the state, farmers lose about $1.5 billion in crops.

1981    More than twenty-one inches of rain falls in less than twenty-four hours in Stephens County in October, resulting in more than $20 million in damages.

1982    Another December storm dumps sixteen inches of snow in Lubbock in twenty-four hours.

1983    Mesquite trees begin to die as a drought continues in the Pecos River valley.

1984    The hottest day of the year occurs August 19 in Glen Rose, 115 degrees.

1985    Delta Flight 191 crashes during a thunderstorm at Dallas-Fort Worth International Airport on August 2, killing 137 passengers and crew members.

1986    More than seven inches of rain is recorded in Athens on April 5.

1987    A May 23 tornado almost wipes out Saragosa, killing twenty-nine and injuring one hundred twenty-one—many of whom were attending a kindergarten graduation at the community hall when the walls collapsed.

1988    Del Rio records a 112-degree day on June 9.

1989    A Christmas Eve cold front sends temperatures plummeting to 15 degrees in the Lower Valley.

1990    Rains plague North Texas from April 29 to May 5, causing flooding along the Trinity, Neches, Sabine, and Brazos Rivers, killing thirteen and causing $1 billion in damages.

1991 The coldest day of the year is recorded at Stratford on November 4, minus 3 degrees.

1992 The hottest day of the year occurs at Castolon on June 17, 117 degrees. Castolon will earn the hottest day honor the following year, although at a cooler 113 degrees.

1993 Stratford again has the coldest day of the year, minus 1 on January 10.

1994 An April 25 tornado kills three and causes major damage in Lancaster, while hail pounds North Texas. The next day another tornado strikes Gainesville, causing major destruction but no deaths.

1995 A May thunderstorm with brutal hail pounds Fort Worth and Dallas, causing $2 billion in damages.

1996 Flash floods November 23-24 cause four deaths in Denton and Tarrant Counties.

1997 An F-5 tornado hits Jarrell on May 27, killing twenty-seven and causing damages estimated at $20 million. According to Baylor University meteorology professor Don Greene, the tornado took 11 to 18 minutes to travel one mile and went only five miles, a short distance for such a powerful tornado.

1998 A heat wave kills one hundred thirty-one and destroys more than $2 million in crops and livestock. The drought is broken by severe flooding in August, September, and October.

1999 Hurricane Bret becomes the biggest hurricane to strike Texas in twenty years when it makes landfall seventy miles south of Corpus Christi on August 22, packing winds of one hundred twenty-five miles per hour and torrential rain.

**Sources**: *Annual Summaries of Weather; The Buffalo Soldiers; The Climates of Texas Counties; Dallas Morning News; The Drought of the 1950s; Fort Concho Medical History 1869 to 1872; Fort Worth Star-Telegram; Late Nineteenth Century Texas Hurricanes; The New Handbook of Texas; 1984 Hurricane Almanac; One Hundred Years of Texas Weather 1880-1979; Santa Fe New Mexican; Significant Floods in the WGRFC Area; Texas Weather; The Tornado; Wave of the Gulf.*

## Appendix B

# 20 Worst Tornadoes in U.S. History

| | Date | Site | Deaths |
|---|---|---|---|
| 1. | March 18, 1925 | Missouri-Illinois-Indiana | 689 |
| 2. | May 6, 1840 | Natchez, Mississippi | 317 |
| 3. | May 27, 1896 | St. Louis, Missouri | 255 |
| 4. | April 5, 1936 | Tupelo, Mississippi | 216 |
| 5. | April 6, 1936 | Gainesville, Georgia | 203 |
| 6. | April 9, 1947 | Woodward, Oklahoma | 181 |
| 7. | April 24, 1908 | Amite, Louisiana/Purvis, Mississippi | 143 |
| 8. | June 12, 1899 | New Richmond, Wisconsin | 117 |
| 9. | June 8, 1953 | Flint, Michigan | 115 |
| **10.** | **May 11, 1953** | **Waco, Texas** | **114** |
| **10.** | **May 18, 1902** | **Goliad, Texas** | **114** |
| 12. | March 23, 1913 | Omaha, Nebraska | 103 |
| 13. | May 26, 1917 | Mattoon, Illinois | 101 |
| 14. | June 23, 1944 | Shinnston, West Virginia | 100 |
| 15. | April 18, 1880 | Marshfield, Missouri | 99 |
| 16. | June 1, 1903 | Gainesville/Holland, Georgia | 98 |
| 16. | May 9, 1927 | Poplar Bluff, Missouri | 98 |
| 18. | May 10, 1905 | Snyder, Oklahoma | 97 |
| 19. | April 24, 1908 | Natchez, Mississippi | 91 |
| 20. | June 9, 1953 | Worcester, Massachusetts | 90 |

**Source**: Storm Prediction Center

## Appendix C

# 20 Deadliest 20ᵗʰ-Century Hurricanes on U.S. Mainland

| No. | Hurricane | Year | Category | Deaths |
|-----|-----------|------|----------|--------|
| 1. | **Galveston, Texas** | **1900** | **4** | **8,000** |
| 2. | Florida | 1928 | 4 | 1,836 |
| 3. | **Florida Keys, South Texas** | **1919** | **4** | **600** |
| 4. | New England | 1938 | 3 | 600 |
| 5. | Florida Keys | 1935 | 5 | 408 |
| 6. | **Audrey (Louisiana, Texas)** | **1957** | **4** | **390** |
| 7. | Northeastern U.S. | 1944 | 3 | 390 |
| 8. | Grand Isle, Louisiana | 1909 | 4 | 350 |
| 9. | New Orleans, Louisiana | 1915 | 4 | 275 |
| 10. | **Galveston, Texas** | **1915** | **4** | **275** |
| 11. | Camille (Mississippi, Louisiana, Virginia) | 1969 | 5 | 256 |
| 12. | Florida, Mississippi, Alabama | 1926 | 4 | 243 |
| 13. | Diane (Northeastern U.S.) | 1955 | 1 | 184 |
| 14. | Southeastern Florida | 1906 | 2 | 164 |
| 15. | Mississippi, Alabama | 1906 | 3 | 134 |
| 16. | Agnes (Florida, NE U.S.) | 1972 | 1 | 122 |
| 17. | Hazel (South Carolina, North Carolina) | 1954 | 4 | 95 |
| 18. | Betsy (Florida, Louisiana) | 1965 | 3 | 75 |
| 19. | Carol (NE U.S.) | 1954 | 3 | 60 |
| 20. | Florida, Louisiana, Mississippi | 1947 | 4 | 51 |

**Source**: National Oceanic and Atmospheric Administration

# Notes

## Chapter 1

1. Primary sources for this chapter are: Merrill, William E., *Captain Benjamin Merrill and the Merrill Family of North Carolina*; Lynch, Dudley, *Tornado Texas Demon in the Wind*; Santerre, George, *White Cliffs of Dallas*; Hutto, Nelson, *The Dallas Story from Buckskins to Top Hat*; Baker, T. Lindsay, *Ghost Towns of Texas*; Bomar, George W., *Texas Weather: The Climates of Texas Counties*; Amerkhan, Ellen, *La Reunion: A Legacy to Dallas* (thesis for Southern Methodist University); *Dallas Weekly Herald*, May 10, 1856; and *Annual Summaries of Weather 1841-1899*, Internet.
2. Merrill, p. 42.
3. Unofficial reports of Texas storms date to at least 1766, but the National Weather Service wasn't established until February 9, 1870. In *One Hundred Years of Texas Weather 1880-1979*, authors John F. Griffiths and Greg Ainsworth write that there were no serious tornadoes in the 1880s. Lynch cites a March 31, 1889, tornado that killed three and injured three near Hamilton.
4. Merrill, pp. 42-43. The *Dallas Weekly Herald* also lists nine killed and twelve injured.
5. The account comes from the *Dallas Weekly Herald*, May 10, 1856.
6. Merrill, p. 43. *Dallas Weekly Herald*, May 10, 1856.
7. "Blue Norther," *The New Handbook of Texas Online*, Internet.
8. Santerre, p. 62.

## Chapter 2

1. Primary sources: *Indianola Scrap Book: Fiftieth Anniversary of the Storm of August 20, 1886*, *The Victoria Advocate*; Malsch, Brownson, *Indianola: The Mother of Western Texas*; Bomar, George W., *Texas Weather*; Baker, T. Linday, *Ghost Towns of Texas*; Ellis, Michael J., *1984 Hurricane Almanac*; Ludlum, David M., *Early American Hurricanes 1492-1870*; Henry, Walter K., Driscoll, Dennis M., and McCormack, J. Patrick, *Hurricanes on the Texas Coast: The Destruction*; Wilkins, Frederick, *The Law Comes to Texas: The Texas Rangers 1870-1901*; *The Deadliest Atlantic Tropical Cyclones, 1492-1994*, NOAA Technical Memorandum; Mrs. Ernestine Weiss Faudie interview, *American Life Histories: Manuscripts from the Federal Writers' Project, 1936-1940*, Internet. "The Indianola Flood," "Hurricane

Devastation of an Earlier Day," Lennie E. Stubblefield papers, King Ranch Archives; *The New Handbook of Texas Online*, Internet; and *Late Nineteenth Century Texas Hurricanes*, Internet.

2. The account comes from *Indianola Scrap Book*, Stubblefield and Malsch.
3. Stubblefield. This comes from a letter from H. Seeligson, who mistakenly calls September 15 a Tuesday. His details of the storm, however, match other accounts. Who was H. Seeligson? A pretty good guess would be Henry Seeligson, a prominent Indianola businessman at the time.
4. Stubblefield.
5. Malsch, p. 239.
6. *Indianola Scrap Book*, p. 75. Malsch, p. 245.
7. The account comes from *Indianola Scrap Book*, Mrs. Ernestine Weiss Faudie interview, and Malsch. According to the *Advocate*, Blackburn was charged with rape and Ruschau with cattle theft.
8. Stubblefield. Malsch, p. 244.
9. Ellis, p. 27.
10. Stubblefield. *Indianola Scrap Book*, p. 76.
11. "Indianola Hurricanes," *The New Handbook of Texas Online*, Internet; *Indianola Scrap Book*, p. 82. Malsch, pp. 237, 251.
12. Malsch, pp. 237-238. Stubblefield.
13. *Indianola Scrap Book*, pp. 79-82.

## Chapter 3

1. After the flood, the name San Angela was rejected by the post office because it was grammatically incorrect. Forced to choose between the correct Santa Angela and San Angelo, residents opted for the latter.
2. In "The Story of Benficklin," *West Texas Historical Association Yearbook* (1946), Mary Bain Spence says the population of Ben Ficklin at the time of the flood was three hundred.
3. Leckie, Shirley Anne, editor, *The Colonel's Lady on the Western Frontier: The Correspondence of Alice Kirk Grierson*, p. 72.
4. *Tom Green Times*, August 26, 1882. In addition to Spence and the *Tom Green Times* (August 26, September 2, September 16, 1882), primary sources on the flood are: Miles, Susan, "Until the Flood 1867-1882," *Edwards Plateau Historian*, Volume 2, 1966; Waring, Katharine T., "Ben Ficklin's Flood," *Fort Concho Report*, Fall 1982; "1882: The Year in Review," *Fort Concho Report*, Winter 1982; "Karger Tells of Flood Loss at Ben Ficklin," *Frontier Times*,

November 1928; Kelton, Elmer, *Elmer Kelton Country: The Short Nonfiction of a Texas Novelist*; Griffiths, John F., and Ainsworth, Greg, *One Hundred Years of Texas Weather 1880-1979*; Mrs. Cicero Russell and Becky Sanford interviews, *American Life Histories: Manuscripts from the Federal Writers' Project, 1936-1940*, Internet; Elmer Kelton interviews; and *The New Handbook of Texas Online*, Internet.

5. Spence, pp. 41-42. Metcalfe told his story to the *San Angelo Standard*. Waring, p. 3.
6. Miles, p. 25.
7. Griffiths, p. 10. "Ben Ficklin," *The New Handbook of Texas Online*. Spence, p. 42. The *Tom Green Times* cited the rainfall at 5.85 inches. The figure used is from Griffiths. The *Times* said the maximum water height was 45 feet. The figure used is from Spence. To put the rainfall in perspective, the National Weather Service in San Angelo reported in 1998 that the average rainfall in August was 1.93 inches.
8. Russell gave her interview to Nellie B. Cox in San Angelo on February 2, 1938.
9. The rescue attempt and aftermath are based on the *Times* report and Waring.
10. *Tom Green Times*, August 26, September 2, 1882.
11. Waring, p. 16.
12. Sanford had been "living not far from Ben Ficklin when it was washed away by the flood." She was living in San Angelo when she was interviewed by Nellie B. Cox on February 10, 1938.
13. Spence, p. 42.
14. Ibid, p. 45.
15. Elmer Kelton interviews.

## Chapter 4

1. Baron von Richthofen, Walter, *Cattle-Raising on the Plains of North America*, pp. 32-34.
2. Ibid, p. 99.
3. Primary sources: Wheeler, David L. "The Texas Panhandle Drift Fences," *Panhandle-Plains Historical Review*, Volume 55, 1982; Wheeler, David L., "The Blizzard of 1886 and Its Effect on the Range Cattle Industry in the Southern Plains," *Southwestern Historical Quarterly*, Volume 94, 1991; Haley, J. Evetts, *Charles Goodnight: Cowman and Plainsman*; Nordyke, Lewis, *Great Roundup: The Story of Texas and Southwestern Cowmen*; McCallum, Henry D. and

Frances, *The Wire That Fenced the West*; *Amarillo News-Globe*, Golden Anniversary Edition, 1938, January 23, 1944; and *The New Handbook of Texas Online*, Internet.
4. "Drift Fences," p. 26.
5. Ibid., p. 31.
6. Quoted in "Drift Fences," p. 34.
7. *Amarillo News-Globe*, January 23, 1944.
8. Quoted in "Blizzard of 1886," p. 425.
9. "Blizzard of 1886," p. 427.
10. *Amarillo News-Globe*, Golden Anniversary Edition, 1938.
11. Ibid., January 23, 1944.
12. The Carter and Peterson accounts come from Nordyke, p. 155.
13. "Blizzard of 1886," pp. 428-429.
14. Ibid., p. 428.
15. Ibid., p. 429.
16. "Drift Fences," p. 35.

**Chapter 5**
1. Mrs. Ernestine Weiss Faudie interview, *American Life Histories: Manuscripts from the Federal Writers' Project, 1936-1940*.
2. Primary sources, in addition to Faudie, are: Malsch, Brownson, *Indianola: The Mother of Western Texas*; *Indianola Scrap Book: Fiftieth Anniversary of the Storm of August 20, 1886*, The Victoria Advocate; Baker, T. Lindsay, *Ghost Towns of Texas*; Ellis, Michael J., *1984 Hurricane Almanac*; Williams, J.W., "A Statistical Study of the Drouth of 1886," *West Texas Historical Association Year Book*, 1945; *The Deadliest Atlantic Tropical Cyclones, 1492-1994*, NOAA Technical Memorandum; *Galveston Daily News*, August 1886 editions; and Internet sources *Upper Texas Coast Tropical Cyclones in the 1880s, Late Nineteenth Century Texas Hurricanes, Annual Summaries of Weather 1841-1899, The New Handbook of Texas Online*, and *The Indianola Hurricane of 1886*.
3. Malsch, p. 262.
4. Faudie. "The Deadliest Atlantic Tropical Cyclones, 1492-1994," p. 24. *The Indianola Hurricane of 1886*. Death figures recorded include 20, 28, and 46. As *The Indianola Hurricane of 1886* Internet site points out, little is known about this storm because so little survived.
5. *Indianola Scrap Book*, pp. 133, 142-143. Malsch, p. 264. *The Indianola Hurricane of 1886*.

6. *Indianola Scrap Book*, p. 134. Ellis, p. 28.
7. *Indianola Scrap Book*, p. 143.
8. Ibid., p. 127.
9. Ibid., p. 144.
10. Ibid., p. 134. Malsch, p. 264.
11. *Indianola Scrap Book*, p. 143.
12. Ibid., p. 130.
13. Ibid., p. 128.
14. *Galveston Daily News*, August 24, 1886. *Indianola Scrap Book*, p. 129.
15. *Indianola Scrap Book*, pp. 135, 145.
16. Ibid., p. 146.
17. Ibid., p. 145.
18. *Galveston Daily News*, August 24, 1886. Malsch, p. 265. *Annual Summaries of Weather 1841-1899*. Ellis, pp. 28-29.
19. *Indianola Scrap Book*, pp. 146-147.
20. Ibid., p. 147.
21. Quoted in Baker, p. 66.

## Chapter 6

1. *Climatic Variability and its Implications for Sustainable Agriculture*, Internet.
2. Primary sources: Alonzo, Armando C., "Change and Continuity in Tejano Ranches in the Trans-Nueces, 1848-1900," *Ranching in South Texas: A Symposium*; Dunn, Roy Sylvan, "Drouth in West Texas, 1890-1894," *West Texas Historical Association Year Book*, 1961; Williams, J.W., "A Statistical Study of the Drouth of 1886," *West Texas Historical Association Year Book*, 1945; O'Neal, Bill, *Historic Ranches of the Old West*; Wallis, George A., *Cattle Kings of the Staked Plains*; Haley, J. Evetts, *Charles Goodnight: Cowman and Plainsman* and *The XIT Ranch of Texas and the Early Days of the Llano Estacado*; Webb, Walter Prescott, *The Great Plains*; Nixon, Jay, *Stewards of a Vision*; Lea, Tom, *The King Ranch*; Griffiths, John F., and Ainsworth, Greg, *One Hundred Years of Texas Weather 1880-1979*; Bentley, Max, "The Disappearing Ranch," untitled article and miscellaneous items, King Ranch Archives; and *The New Handbook of Texas Online*, Internet.
3. Dunn, pp. 130, 133. Lea, p. 484.
4. Kelton, Elmer, *The Time It Never Rained*, p. x.
5. Dunn, p. 129.

6. Ibid., p. 130.
7. Ibid., p. 132.
8. *Charles Goodnight*, p. 383.
9. Fehrenbach, T.R., *Lone Star: A History of Texas and the Texans*, p. 611.
10. Lea, p. 502.
11. Untitled article, King Ranch Archives.
12. Ibid., The author gives the date as 1893, but the first artesian well didn't produce until 1899. This could be the third well, and the year 1903.
13. Bentley.
14. Webb, pp. 376-377.
15. Griffiths, p. 29.

**Chapter 7**

1. O'Neal, Bill, *Historic Ranches of the Old West*, p. viii.
2. Richard King Letters, King Ranch Archives.
3. Unless otherwise noted, information for this chapter comes from "Experiments in Production of Rainfall," Senate Executive Document No. 45, 52nd Congress, First Session, National Archives.
4. Spence, Clark C., *The Rainmakers*, p. 30.
5. Ibid., pp. 33-34.
6. Rainmaking Report, p. 58. Lea, Tom, *The King Ranch*, p. 503.

**Chapter 8**

1. Hoyt, William G., and Langbein, Walter B., *Floods*, p. 296.
2. Primary sources: *The Handbook of Waco and McLennan County, Texas*; Baker, J.W., A History of Robertson County, Texas; McCarver, Norman L., and McCarver, Norman L. Jr., Hearne on the Brazos; Sowell, A.J., *A History of Fort Bend County*; *Mennonitische Rundschau*, July 12, July 19, 1899; *Flood Plain Information: Brazos River, Fort Bend County, Texas,* U.S. Army Corps of Engineers Galveston District publication, February 1977; Kathy Stracener letter; *Climate and Crops: Texas Selection*, June 1899; and Internet sources *The New Handbook of Texas Online, Significant Texas Floods*, and *Hearne, Texas History*.
3. *Mennonitische Rundschau*, July 12, 1899. All letters from the *Mennonitische Rundschau* in this chapter were translated from German by W.M. Von-Maszewski, department manager, genealogy/local history, George Memorial Library, Richmond, Texas.
4. Ibid.

5. Sowell, pp. 333-334.
6. *Mennonitische Rundschau*, July 12, 1899.
7. Ibid., July 19, 1899.
8. Climate and Crops, p. 4. *Significant Texas Floods*.

## Chapter 9

1. *Galveston Daily News*, September 13, 1900.
2. Primary sources: Weems, John Edward, *A Weekend in September*; Cartwright, Gary, *Galveston: A History of the Island*; Mason Jr., Herbert Molloy, *Death from the Sea: Our Greatest Natural Disaster: The Galveston Hurricane of 1900*; Ousley, Clarence, editor, *Galveston in Nineteen Hundred*; Hoyt, William G., and Langbein, Walter B., *Floods*; Coulter, John, editor, *The Complete Story of the Galveston Horror*; Halstead, Murat, *Galveston: The Horrors of a Stricken City*; Sanders, Ti, *Weather: A User's Guide to the Atmosphere*; Taft, Harold, and Godbey, Ron, *Texas Weather*; Bomar, George W., *Texas Weather; The Deadliest, Costliest, and Most Intense United States Hurricanes of This Century*, NOAA Technical Memorandum; *Some Devastating North Atlantic Hurricanes of the 20th Century*, U.S. Department of Commerce, National Oceanic and Atmospheric Administration; Lloyd Fayling Papers, Anonymous letter, and John D. Blagden Papers, Rosenberg Library archives; *Kalamazoo Daily Telegraph*, September 20, 1900; *Los Angeles Daily Herald*, September 13, 1900; *Galveston Daily News*, September 1900 editions; and Internet sources *The New Handbook of Texas Online*, "The Galveston Hurricane - The Last and Worst Hurricane of the 19th Century," "Upper Texas Coast Tropical Cyclones in the 1900s," and "The 1900 Storm: Sisters of Charity Orphanage."
3. Mason, p. 68.
4. Ibid., p. 90.
5. Ibid., p. 88.
6. Lloyd Fayling Papers.
7. *Galveston Daily News*, September 14, 1900.
8. Ibid.
9. Ibid., The torn copy of the newspaper, on microfilm at the Rosenberg Library in Galveston, is hard to read in places.
10. Anonymous letter, Rosenberg Library archives.
11. Lloyd Fayling Papers.
12. Ousley, p. 116. Cartwright, p. 167. Weems, p. 93.
13. *Galveston Daily News*, September 14, 1900.

14. The Fayling account is taken from the Lloyd Fayling Papers, with additional information from Mason and *Kalamazoo Daily Telegraph*, September 20, 1900.
15. *Los Angeles Daily Herald*, September 13, 1900.
16. John D. Blagden Papers.
17. Ousley, p. 117.
18. Lloyd Fayling Papers.

**Chapter 10**

1. Lloyd Fayling Papers. For primary sources used in this chapter, see footnote 2 for Chapter 9. Additional main sources are: *Kalamazoo Daily Telegraph*, October 17, 1900; Report of Clara Barton, and Fred Napp Papers, Rosenberg Library archives.
2. Lloyd Fayling Papers.
3. John D. Blagden Papers.
4. *Kalamazoo Daily Telegraph*, September 20, 1900.
5. Coulter, p. 57.
6. Ibid., p. 45.
7. Ibid., p. 85.
8. Halstead, p. 109.
9. Cartwright, p. 175. Ousley, p. 116.
10. Ousley, p. 281.
11. *Kalamazoo Daily Telegraph*, October 17, 1900.
12. John D. Blagden Papers.
13. *Galveston Daily News*, September 12, 1900.
14. Ousley, p. 117.
15. Ibid.
16. Mason, p. 199.
17. Lloyd Fayling Papers.
18. Report of Clara Barton.
19. Coulter, p. 108.
20. Lloyd Fayling Papers. Coulter, p. 58.
21. Coulter, p. 56.
22. John D. Blagden Papers.
23. Clara Barton Report.
24. *Los Angeles Daily Herald*, September 13, 1900.
25. Fred Napp Papers.
26. *Los Angeles Daily Herald*, September 13, 1900.
27. Coulter, pp. 65-66.

28. Ousley, p. 280.
29. The Ward and Barton accounts come from Clara Barton Report.
30. Coulter, p. 69.
31. Ousley, pp. 216, 246.
32. *Galveston Daily News*, September 16, 1900.
33. Ibid.

**Chapter 11**

1. Primary sources: Pruett, Jakie L., and Cole, Everett B., *The History & Heritage of Goliad County*; White, Beth, *Goliad Remembered 1836-1940*; *San Antonio Semi-Weekly Express*, May 23, 1902; *Victoria Weekly Advocate*, May 24, 1902, May 15, 1977, and undated 1992 clipping, Goliad Library; "The Goliad Tornado," *Climate and Crops - Texas Section*, May 1902; Doris Freer letter; and *Annual Summaries of Weather 1841-1899* and *The New Handbook of Texas Online*, Internet.
2. *Goliad County*, p. 39. White, pp. 63-64.
3. White, p. 64. White cites the account that had been reprinted in the May 11, 1922, *Goliad Advance-Guard*.
4. *Victoria Weekly Advocate*, May 24, 1902.
5. "Tornadoes," *The New Handbook of Texas Online*. *Victoria Advocate*, May 15, 1977. White, p. 64. There are many similarities between the *Guard* account and Browne's version told to the *Advocate*.
6. *Victoria Advocate*, undated 1992 clipping, Goliad Library.
7. *Victoria Advocate*, May 15, 1977. *The History & Heritage of Goliad County* gives Kate's age at 10 and Warren's at 3 during the storm.
8. *Goliad County*, p. 40.
9. *Victoria Advocate*, May 15, 1977. *Goliad County*, pp. 39-40.
10. *Goliad County*, p. 40.
11. White, p. 64. *Victoria Advocate*, undated 1992 clipping. *San Antonio Semi-Weekly Express*, May 23, 1902. A witness told the *Express* the tornado lasted only two or three minutes.
12. "The Goliad Tornado," p. 4.
13. *Victoria Advocate*, undated 1992 clipping.
14. *Goliad County*, p. 40. *San Antonio Semi-Weekly Express*, May 23, 1902.
15. White, pp. 65-66.
16. Ibid., pp. 66-67.
17. *Victoria Advocate*, May 15, 1977.
18. White, pp. 66-67. *Victoria Weekly Advocate*, May 24, 1902.

19. *Victoria Advocate*, May 15, 1977, and undated 1992 clipping.
20. *San Antonio Semi-Weekly Express*, May 23, 1902.
21. White, p. 70. *San Antonio Semi-Weekly Express*, May 23, 1902.
22. White, p. 71.

## Chapter 12

1. The Fuller account comes from Fuller, Theodore A., *When the Century and I Were Young*, Corpus Christi Public Library. Other primary sources are: *Corpus Christi Caller-Times* (October 18, 1955; June 18, 1959; September 13, 1959; September 11, 1960; September 18, 1961; June 2, 1978; May 3, 1979; August 21, 1980; August 19, 1984; June 3, 1987; August 31, 1991); undated 1919 newspaper clipping, Corpus Christi Public Library; J.H.C. White telegram to Mrs. J. G. Kenedy, August 19, 1916, King Ranch Archives; Johnson, Harriett, *Hurricanes; Centennial History of Corpus Christi; Manuscripts for a History of Nueces County*, Corpus Christi Public Library; *The Deadliest, Costliest, and Most Intense United States Hurricanes of This Century*, NOAA Technical Memorandum; and *The New Handbook of Texas Online*, Internet.
2. Quoted in Johnson, pp. 2-3.
3. *Corpus Christi Caller-Times*, September 13, 1959. (Copyright 1999 Caller-Times Publishing Company. Reprinted with permission. All rights reserved.)
4. Ibid., August 19, 1984.
5. Undated 1919 newspaper clipping, Corpus Christi Public Library.
6. *Corpus Christi Caller-Times*, September 13, 1959. (Copyright 1999 Caller-Times Publishing Company. Reprinted with permission. All rights reserved.) *Centennial History of Corpus Christi*, p. 117.
7. *Corpus Christi Caller-Times*, September 13, 1959. (Copyright 1999 Caller-Times Publishing Company. Reprinted with permission. All rights reserved.)
8. Ibid.
9. Johnson, p. 10.
10. *Manuscripts for a History of Nueces County. Corpus Christi Caller-Times*, September 11, 1960. (Copyright 1999 Caller-Times Publishing Company. Reprinted with permission. All rights reserved.)
11. Ibid.
12. Ibid., September 11, 1960 and October 18, 1955. *Manuscripts for a History of Nueces County*, p. 237.

13. Undated 1919 newspaper clipping, Corpus Christi Public Library.
14. Johnson, p. 12.
15. *Corpus Christi Caller-Times*, June 18, 1959. (Copyright 1999 Caller-Times Publishing Company. Reprinted with permission. All rights reserved.)
16. Ibid., August 21, 1980.
17. Ibid., June 2, 1978.
18. Ibid.
19. Ibid.
20. Johnson, p. 14.
21. Ibid., p. 14. *Corpus Christi Caller-Times*, June 2, 1978. (Copyright 1999 Caller-Times Publishing Company. Reprinted with permission. All rights reserved.)
22. Ibid., September 18, 1961.

**Chapter 13**

1. Primary sources: *Austin American*, February 18, 1960; *Austin Statesman* (April 14, 1954; August 28, 1960; June 14, 1970); *Austin American-Statesman*, May 4, 1958; Undated *Austin Statesman* clipping, Austin History Center, Austin Public Library; Untitled notes, Austin History Center, Austin Public Library. Simonds, Frederic W., "The Austin, Texas Tornadoes of May 4, 1922," *University of Texas Bulletin*, February 15, 1923; and *The New Handbook of Texas Online*.
2. *Austin American*, February 18, 1960.
3. *Austin Statesman*, August 28, 1960.
4. Ibid., April 14, 1954.
5. *Austin American*, February 18, 1960.
6. Simonds, p. 1.
7. *Austin American-Statesman*, May 4, 1958.
8. *Austin Statesman*, April 14, 1954.
9. *Austin American-Statesman*, May 4, 1958.
10. Ibid.
11. Ibid.
12. Ibid.
13. Untitled notes, Austin History Center, Austin Public Library.
14. *Austin Statesman*, August 28, 1960.
15. Undated *Austin Statesman* clipping, Austin History Center, Austin Public Library.

## Chapter 14

1. Primary sources: *Abilene Morning News*, May 7-8, 1930; *Austin American*, May 7-8, 1930; *Austin Statesman*, May 6, 1930; *Gonzales Inquirer*, May 8, 15, 1930; *Kenedy Advance*, May 8, 1930; *San Antonio Light*, May 7, 1930; *San Antonio Express*, May 7, 1930; Bomar, George W., *Texas Weather*; *Climatological Data*, Texas Section, General Summary, Houston, Texas, U.S. Department of Agriculture Weather Bureau, May 1930; and *The New Handbook of Texas Online*, Internet.
2. *Austin Statesman*, May 6, 1930. *Abilene Morning News*, May 7, 1930. "Tornadoes," *The New Handbook of Texas Online*.
3. *Abilene Morning News*, May 7, 1930. Climatological Data cites Abilene's maximum velocity at fifty-one miles per hour.
4. *Gonzales Inquirer*, May 8, 1930. *Austin American*, May 7, 1930. The population of Frost at the time of the tornadoes is given as five hundred to one thousand.
5. *Gonzales Inquirer*, May 8, 1930. *San Antonio Light*, May 7, 1930.
6. *Gonzales Inquirer*, May 8, 1930.
7. *San Antonio Light*, May 7, 1930. *Austin American*, May 7, 1930. *San Antonio Express*, May 7, 1930. *Gonzales Inquirer*, May 8, 1930.
8. *Gonzales Inquirer*, May 8, 1930.
9. "Tornadoes." *San Antonio Light*, May 7, 1930. *Gonzales Inquirer*, May 8, 1930.
10. *San Antonio Light*, May 7, 1930.
11. *San Antonio Express*, May 7, 1930. *San Antonio Light*, May 7, 1930. *Kenedy Advance*, May 8, 1930.
12. *Gonzales Inquirer*, May 8, 1930.
13. *San Antonio Light*, May 7, 1930.
14. *Austin American*, May 8, 1930. *Gonzales Inquirer*, May 15, 1930.

## Chapter 15

1. Primary sources: *Amarillo Daily News*, March 4, April 15, 1935; *Amarillo News-Globe*, March 11, 1962; *Amarillo Globe-News*, April 13, 1975; *Lubbock Evening Journal*, April 15, 1935; *Ochiltree County Herald*, April 18, 1935; Max Evans, J.T. McLarty, and Richard and Rita Sell interviews; Sell, Richard, "Black Sunday 1935" (unpublished manuscript, courtesy of Richard Sell); and *Black Sunday* and *The New Handbook of Texas Online*, Internet.
2. *Lubbock Evening Journal*, April 15, 1935. Max Evans, J.T. McLarty interviews. *Amarillo Daily News*, March 4, 1935. *Amarillo*

*News-Globe*, March 11, 1962. "Dust Bowl," *The New Handbook of Texas Online.*

3. Sell, pp. 3-4.
4. The Sell account comes from Sell, Richard, "Black Sunday 1935."
5. J.T. McLarty interview.
6. The McLarty account comes from a 1999 interview with J.T. McLarty.
7. *Amarillo Daily News*, April 15, 1935.
8. J.T. McLarty interview. *Ochiltree County Herald*, April 18, 1935.

**Chapter 16**

1. Primary sources: Elmer Kelton, C.F. Eckhart, Jay O'Brien, Juliette Forchheimer Schwab interviews; Margaret Farley letter; *Albuquerque Journal*, December 8, 1998; *Wall Street Journal*, November 8, 1952; "Nature Control Is a Business," *Business Week*, August 5, 1950; Fort Stockton Pioneer, 1954-56 editions; *Sanderson Times*, 1950-54 editions; Kingsville Chamber of Commerce resolution, January 2, 1951, King Ranch Archives; V.W. Lehmann memorandum to R.J. Kleberg Jr., March 18, 1951, King Ranch Archives; Ford Hubbard letter to Robert J. Kleberg, April 12, 1951, King Ranch Archives; Robert J. Kleberg Jr. letter to Emil A. Hanslin, King Ranch Archives; and *The Drought of the 1950s* and *The New Handbook of Texas Online*, Internet.
2. Elmer Kelton interviews. *Sanderson Times*, March 30, 1951, July 27, 1951. "Droughts," *The New Handbook of Texas Online.*
3. V.W. Lehmann memorandum to R.J. Kleberg Jr., March 18, 1951, King Ranch Archives.
4. *Wall Street Journal*, November 8, 1952. "Nature Control Is a Business," *Business Week*, August 5, 1950. Ford Hubbard letter to Robert J. Kleberg, April 12, 1951, King Ranch Archives.
5. Robert J. Kleberg Jr. letter to Emil A. Hanslin, King Ranch Archives.
6. *Sanderson Times*, November 28, 1952.
7. *The Drought of the 1950s.*

**Chapter 17**

1. Primary sources for the San Angelo tornado: Elmer Kelton interviews; *Dallas Morning News*, May 12-13, 1953; *Dallas Times Herald*, May 12, 1953. Moore, Harry Estill, *Tornadoes Over Texas: A Study of Waco and San Angelo in Disaster*; Texas-May 1953, E.A. Farrell, Section Director - Houston, U.S. Department of Agriculture Weather Bureau.

2. Primary sources for the Waco tornado: *Waco Tornado 1953: Force that Changed the Face of Waco*, Volumes 1-2; Weems, John Edward, *The Tornado*; Poage, W.R. (Bob), *McLennan County—Before 1980*; *Dallas Times Herald*, May 12-13, 1953; *Dallas Morning News* (May 12, 1953; May 11, 1963); "Waco Disaster, The," *Waco Heritage & History*, Spring 1981; *Waco News-Tribune*, May 12-16, 1953; *Waco Times-Herald*, May 12-16, 1963; *Waco Record*, April 10, 1953; *The 1953 Waco Tornado: Tragedy and Triumph* (video).
3. Carl M. Barrett interview, *Waco Tornado 1953: Force that Changed the Face of Waco*, Volume 1. This two-volume 1981 edition from the Waco-McLennan County Library contains interviews of storm survivors. Unless otherwise noted, quoted recollections of the Waco tornado in this chapter come from these two volumes.
4. *Dallas Times Herald*, May 12, 1953. Dr. H. Joe Jaworski interview, *Waco Tornado 1953*, Volume 1.
5. *Waco News-Tribune*, May 14, 1953. *Dallas Times Herald*, May 12, 1953.
6. *Waco Times Herald*, May 15, 1953.

**Chapter 18**

1. Primary sources: Scogin, Russell Ashton, *The Sanderson Flood of 1965: Crisis in a Rural Texas Community*; *San Angelo Standard-Times* (June 12-13, June 18, August 29, 1965; June 11, 1967; April 28, 1998); *Fort Stockton Pioneer*, June 17, 1965; *Sanderson Times*, June 18, 1965; *El Paso Times*, June 12, 1965; *Southern Pacific Bulletin*, July 1965; Frances Corbett letter; Susan Corbett interviews; Bomar, George W., *Texas Weather*; *The Climates of Texas Counties*; and *The New Handbook of Texas Online*, Internet.
2. Frances Corbett letter.
3. *San Angelo Standard-Times*, August 28, 1998. Reprinted by permission of the *San Angelo Standard-Times*.
4. *Fort Stockton Pioneer*, June 17, 1965. The newspaper gave a chronological account of the flood, saying "we are sparing our readers the more sensational details which have been thoroughly covered by the daily press. . . ."
5. Frances Corbett letter.
6. Ibid.
7. *San Angelo Standard-Times*, June 11, 1967. Reprinted by permission of the *San Angelo Standard-Times*.
8. Frances Corbett letter. Susan Corbett interviews. Scogin, p. 33.
9. Frances Corbett letter.

10. Ibid., Susan Corbett interviews. Scogin, pp. 77-78.
11. *San Angelo Standard-Times*, June 13, 1965. Reprinted by permission of the *San Angelo Standard-Times*.
12. Scogin, pp. 64, 84. *San Angelo Standard-Times*, June 12-13, 1965. Reprinted by permission of the *San Angelo Standard-Times*.
13. Ibid., June 13, 1965.
14. Ibid.
15. Ibid., June 12, 1965.
16. Ibid. Scogin, p. 84. The *Standard-Times* listed Nicholas Flores as eighty-two years old and Mrs. Flores as eighty. Scogin, the *El Paso Times* and *Fort Stockton Pioneer* give Nicholas Flores's age as seventy.
17. *San Angelo Standard-Times*, June 12, 1965. Reprinted by permission of the *San Angelo Standard-Times*.
18. Ibid., June 11, 1967.
19. Frances Corbett letter.
20. *San Angelo Standard-Times*, June 13, 1965. Reprinted by permission of the *San Angelo Standard-Times*.
21. Ibid., June 11, 1967.
22. Scogin, p. 40. *Sanderson Times*, June 18, 1965.
23. Frances Corbett letter.
24. Ibid.
25. *San Angelo Standard-Times*, June 12, 1965. Reprinted by permission of the *San Angelo Standard-Times*.
26. Ibid.
27. Susan Corbett interviews.
28. *San Angelo Standard-Times*, August 29, 1965. Reprinted by permission of the *San Angelo Standard-Times*.
29. Ibid., April 28, 1998.

**Chapter 19**
1. Cynthia Bush account, Moffett Library, Midwestern State University. Other primary sources: *Dallas Times Herald*, April 11-12, 1979; *Dallas Morning News*, April 11-12, 1979; *Wichita Falls Record News*, April 12-17, 1979; *Wichita Falls Times*, April 11-17, 1979; *Dallas Times Herald-Wichita Falls Times*, "joint community effort," April 11, 1979; *The Wichita Falls, Vernon & Lawton Tornadoes*; untitled tornado report, Moffett Library; Ann Bryant, Gary Hardee, Mike Dougherty, and Charles Clines interviews; and *The New Handbook of Texas Online*, Internet.

2. The Bush information comes from Cynthia Bush account, Moffett Library, Midwestern State University. Some background is from *Dallas Times Herald*, April 11, 1979.

3. The Bryant account comes from a 1999 interview with Ann Bryant.

4. Background is from *Dallas Times Herald*, April 12, 1979.

5. Charles Clines interview.

6. *Dallas Times Herald-Wichita Falls Times*, "joint community effort," April 11, 1979.

7. *Wichita Falls Record News*, April 12-13,1979. *The Wichita Falls, Vernon & Lawton Tornadoes*, p. 18.

8. *Wichita Falls Times*, April 12, 1979.

9. Cynthia Bush account.

## Chapter 20

1. B.R. Begay letter.

2. Primary sources: *Saragosa, Texas, Tornado, May 22, 1987: An Evaluation of the Warning System*; *Dallas Times Herald*, May 23, 25, 27, 1987; *Dallas Morning News*, May 23, 28, 1987; and *1987 Saragosa, Texas Tornado, The*, ("The Saragosa, Texas Tornado, May 22, 1987" by Bill Alexander, Lubbock WSFO, *Storm Track*, September 30, 1987) and *The New Handbook of Texas Online*, Internet.

3. *Dallas Times Herald*, May 25, 1987.

4. Ibid.

5. Ibid., May 23, 1987. *Dallas Morning News*, May 23, 1987, May 25, 1987.

6. *Dallas Times Herald*, May 25, 1987.

7. Ibid.

8. Ibid., May 27, 1987.

## Chapter 21

1. *Presidio County News*, May 31, 1881.

2. *Tascosa Pioneer*, June 26, 1886. Quoted in *Maverick Town: The Story of Old Tascosa* by John L. McCarty, p. 208.

3. The Cummings account comes from a 1999 interview with Tommy Cummings.

4. Steve Kaye interview.

5. Primary sources: *Fort Worth Star-Telegram*, May 2, 1999, April 28, 1996, March 24. 1996, May 7, 1995.

## Chapter 22

1. Dan Langendorf interview.

2. In addition to personal interviews, primary sources are: *Fort Worth Star-Telegram*, September 6, December 27, December 31, 1998, May 13, 1999; *Dallas Morning News*, May 7, July 13, August 5, September 4, 1998; National Oceanic and Atmospheric Administration news release, October 14, 1998; Texas A&M Agriculture Program news releases, August 5, August 19, September 15, 1998.

3. Kurt Iverson interview.

4. Jay O'Brien interview.

5. Elmer Kelton interviews.

## Epilogue

1. Juliette Forchheimer Schwab, Doctor John Key interviews.

2. Charles Clines interview.

3. Dewees, William R., *Letters from an Early Settler of Texas*, pp. 237-238.

4. "Belle Little interview," *American Life Histories: Manuscripts from the Federal Writers' Project, 1936-1940*.

# Bibliography

**Internet**

*Annual Summaries of Weather,*
    http://www.srh.noaa.gov/FTPROOT/FWD/wx-sumear.html
*Background Information,*
    http://twri.tamu.edu/twripubs/WtrResrc/v22n2/text-1.html
*Black Sunday,*
    http://www.mindspring.com/~jwar/dust/black.htm
*Climatic Variability and its Implications for Sustainable*
    *Agriculture, Progress Report, April 14, 1998,*
    http://uregina.ca/~sauchyn/urdendro/progress.html
*Does Weather Modification Really Work?*
    http://twri.tamu.edu/twripubs/WtrResrc/v20n2/text.html
*Drought of the 1950s, The,*
    http://www.twri.tamu.edu/twripubs/WtrResrc/v22n2/
    text-3.html
*Hearne, Texas History,*
    http://www.rtis.com/reg/hearne/ent/historl.htm
*Indianola Hurricane of 1886, The,*
    http://www.srh.noaa.gov/CRP/does/tropics/history/
    1886-5.html
*Late Nineteenth Century Texas Hurricanes,*
    http://www.srh.noaa.gov/FTPROOT/LCH/txlate19hur.htm
*1987 Saragosa, Texas Tornado, The,*
    http://www.storm-track.com/saragosa.htm
    ("The Saragosa, Texas Tornado, May 22, 1987" by Bill Alexander,
    Lubbock WSFO, *Storm Track*, September 30, 1987)
*New Handbook of Texas Online, The,*
    http://www.tsha.utexas.edu/cgi-bin
*Potential of Cloud Seeding for Augmenting Rainwater in Semi-Arid West*
    *Texas, The,*
    http://www.lib.ttu.edu/playa/text94/playa24.htm
*Significant Floods in the WGRFC Area,*
    http://www.noaa.gov/wgrfc/significant_floods.html
*San Angelo Chamber of Commerce,*
    http://sanangelo-tx.com

*Sisters of Charity Orphanage, The,*
  . http://www.1900storm.com/orphanage.html
*25 Deadliest U.S. Tornadoes, The,*
    http://www.spc.noaa.gov/archive/tornadoes/t-deadly.html
*Upper Texas Coast Tropical Cyclones in the 1880s,*
    http://www.srh.noaa.gov/hgx/hurricanes/1880s.htm
*Upper Texas Coast Tropical Cyclones in the 1900s,*
    http://www.srh.noaa.gov/hgx/hurricanes/1900s.htm
*Why Droughts Plague Texas,*
    http://twri.tamu.edu/twripubs/WtrResrc/v22n2/text-0.html

## Videos

*The 1953 Waco Tornado: Tragedy and Triumph,* Impact
    Productions, Waco Chamber of Commerce Community Develop-
    ment Foundation, 1992.

## Private Letters

B.R. Begay to Johnny D. Boggs, July 2, 1999.

Doris Freer to Johnny D. Boggs, May 22, 1999.

Frances Corbett to Susan Corbett, May 28, 1999 (courtesy of Susan
    Corbett).

Kathy Stracener to Johnny D. Boggs, June 17, 1999.

## Interviews

Ann Bryant (by Mike Dougherty), Mansfield, Texas, July 19, 1999.

Charles Clines (by Johnny D. Boggs), by e-mail, May 15, 1999.

Susan Corbett (by Johnny D. Boggs), by e-mail, May 20, 1999, June 4,
    1999.

Tommy Cummings (by Johnny D. Boggs), by e-mail, July 8, 1999.

Mike Dougherty (by Johnny D. Boggs), by e-mail, August 12, 1999.

C.F. Eckhart (by Johnny D. Boggs), Rapid City, South Dakota, July 1,
    1999.

Max Evans (by Johnny D. Boggs), Albuquerque, N.M., April 19, 1999.

Gary Hardee (by Johnny D. Boggs), by e-mail, July 6, 1999.

Kurt Iverson (by Johnny D. Boggs), by e-mail, August 12, 1999.

Steve Kaye (by Johnny D. Boggs), by e-mail, June 7, 1999.

Elmer Kelton (by Johnny D. Boggs), Rapid City, South Dakota, June 30,
    1999; San Angelo, Texas, July 29, 1999.

Doctor John R. Key (by Johnny D. Boggs), Toyahvale, Texas, June 24,
    1999.

Dan Langendorf (by Johnny D. Boggs), by e-mail, June 7, 1999.

J.T. McLarty (by Johnny D. Boggs), by telephone, August 11, 1999.

Jay O'Brien (by Johnny D. Boggs), by e-mail, June 30, 1999.

Juliette Forchheimer Schwab (by Johnny D. Boggs), Alpine, Texas, June 24, 1999.

Richard and Rita Sell (by Johnny D. Boggs), by telephone, August 23, 1999.

**Government, Archival, and Miscellaneous Documents**

Alonzo, Armando C., "Change and Continuity in Tejano Ranches in the Trans-Nueces, 1848-1900," *Ranching in South Texas: A Symposium*, edited by Joe S. Graham, Texas A&M University- Kingsville, 1994.

*American Life Histories: Manuscripts from the Federal Writers' Project, 1936-1940*: Interviews with Mrs. Ernestine Weiss Faudie, Spence Hardie, Belle Little, Mrs. Cicero Russell, and Becky Sanford.

Amerkhan, Ellen, *Reunion: A Legacy to Dallas*, thesis for Southern Methodist University, 1986, Dallas Public Library.

Anonymous Letter, 22-0045, Galveston & Texas History Center, Rosenberg Library, Galveston.

Bentley, Max, "The Disappearing Ranch," undated article, King Ranch Archives.

*C.W. Post: The Man—The Legend.* 100th Year Anniversary, edited by "Unique" Marketing & Design (Post Commerce and Tourism Bureau).

*Climatological Data*, Texas Section, General Summary, Houston, Texas, U.S. Department of Agriculture Weather Bureau, May 1930.

Cynthia Bush paper, Moffett Library, Midwestern State University, Wichita Falls.

*Deadliest Atlantic Tropical Cyclones, 1492-1994, The*, National Oceanic and Atmospheric Administration Technical Memorandum NWS NHC-47, January 1995.

*Deadliest, Costliest, and Most Intense United States Hurricanes of This Century (and Other Frequently Requested Hurricane Facts), The*, National Oceanic and Atmospheric Administration Technical Memorandum NWS TPC-1, February 1997.

"Drought Chops Forests by $342 Million, Officials Say," Texas A&M University Agriculture Program press release, September 15, 1998.

"Drought May Impact Livestock Industry for Years to Come," Texas A&M University Agriculture Program press release, August 5, 1998.

"Droughts More Severe Than Dust Bowl Likely, NOAA Reports," National Oceanic and Atmospheric Administration press release, December 15, 1998.

"Drought Worse Than '96; Cotton Crop's One of Worst Ever," Texas A&M University Agriculture Program press release, August 19, 1998.

"80-Year Record of Rainfall at South Texas Points," King Ranch Archives.

"Experiments in Production of Rainfall," Publications of the U.S. Government, Senate Executive Document 45, 52nd Congress, 1st Session, National Archives and Records Administration.

F. Napp Papers, 76-0026, Galveston & Texas History Center, Rosenberg Library, Galveston.

"Facts Related by Participants in the 1919 Hurricane," compiled by Sister M. Xavier, *Manuscripts for a History of Nueces County*, Nueces County Historical Society, Volume II, 1966, Corpus Christi Public Library.

"Flood Hits," *Southern Pacific Bulletin*, July 1965 (Terrell County Historical Association).

"Flood Plain Information: Brazos River, Fort Bend County, Texas," U.S. Army Corps of Engineers Galveston District publication, February 1977.

Fuller, Theodore A., *When the Century and I Were Young*, Corpus Christi Public Library.

"Goliad Tornado, The," *Climate and Crops - Texas Section*, U.S. Department of Agriculture Weather Bureau, May 1902.

Graf, Le Roy P., *The Economic History of the Lower Rio Grande Valley 1820-1875*, thesis for Harvard University, 1942, King Ranch Archives.

Hubbard, Ford, letter to Robert J. Kleberg Jr., April 12, 1951, King Ranch Archives.

JCH White letters, King Ranch Archives.

J.D. Blagden Papers, 46-0006, Galveston & Texas History Center, Rosenberg Library, Galveston.

Kilgore, Dan, *Corpus Christi: 1900-1930*, Corpus Christi Public Library.

"King Ranch Inc.," King Ranch Archives.

Richard King Letters, King Ranch Archives.

Kleberg Jr., Robert J., letter to Emil A. Hanslin, April 23, 1952, King Ranch Archives.

Kleberg Jr., Robert J., letter to Wallace E. Howell, September 26, 1967, King Ranch Archives.

Lennie E. Stubblefield papers, courtesy of Maggie Maranto, Charlottesville, Virginia, King Ranch Archives.

L.R. Fayling Papers, 80-0021, Galveston & Texas History Center, Rosenberg Library, Galveston.

Memorandum to R.J. Kleberg Jr. from V.W. Lehmann, March 18, 1951, King Ranch Archives.

"Nature Control Is a Business," reprinted from *Business Week*, Aug. 5, 1950, King Ranch Archives.

Sell, Richard, "Black Sunday 1935," unpublished manuscript, courtesy of Richard Sell, Perryton, Texas.

Red Cross Records, 05-0007, Galveston & Texas History Center, Rosenberg Library, Galveston.

"September 1998 Was Warmest on Record—Globally and in United States, NOAA Reports," National Oceanic and Atmospheric Administration press release, October 14, 1998.

*Some Devastating North Atlantic Hurricanes of the 20th Century*, U.S. Department of Commerce, National Oceanic and Atmospheric Administration, December 1994.

*Texas-May 1953*, E.A. Farrell, Section Director - Houston, U.S. Department of Agriculture Weather Bureau.

*Tropical Cyclones of the North Atlantic Ocean, 1871-1992*, Historical Climatology Series 6-2, U.S. Department of Commerce, National Oceanic and Atmospheric Administration, November 1993.

Untitled paper, Moffett Library, Midwestern State University, Wichita Falls.

## Newspapers

*Abilene Morning News*
*Albuquerque Journal*
*Amarillo Daily News*
*Amarillo Globe-News, News-Globe*
*Austin American*
*Austin Statesman*
*Austin American-Statesman*
*Corpus Christi Caller-Times*
*Dallas Morning News*
*Dallas Times Herald*
*Dallas Weekly Herald*
*El Paso Times*
*Fort Stockton Pioneer*
*Fort Worth Star-Telegram*
*Galveston Daily News*
*Galveston Tribune*

*Gonzales Inquirer*
*Houston Daily Post*
*Kalamazoo Daily Telegraph*
*Kenedy Advance*
*Kingsville Record*
*Los Angeles Daily News*
*Lubbock Evening Journal*
*Mennonitische Rundschau*
*Ochiltree County Herald*
*Presidio County News*
*Sanderson Times*
*San Angelo Standard-Times*
*San Antonio Express*
*San Antonio Light*
*San Antonio Semi-Weekly Express*
*Santa Fe New Mexican*
*Tom Green Times*
*Victoria Advocate*
*Victoria Weekly Advocate*
*Waco Times-Herald*
*Waco Tribune-Herald*
*Waco News-Tribune*
*Waco Record*
*Wall Street Journal*
*Wichita Falls Record News*
*Wichita Falls Times*

**Magazine Articles**

Dunn, Roy Sylvan, "Drouth in West Texas, 1890-1894," *West Texas Historical Association Year Book*, 1961.

"1882: The Year in Review," *Fort Concho Report*, Winter 1982.

"Karger Tells of Flood Loss at Ben Ficklin," *Frontier Times*, November 1928.

Miles, Susan, "Until the Flood 1867-1882," *The Edwards Plateau Historian*, Volume 2, 1966.

Spence, Mary Bain, "The Story of Benficklin, First County Seat of Tom Green County, Texas," *West Texas Historical Association Year Book*, 1946.

Simonds, Frederic W., "The Austin, Texas, Tornadoes of May 4, 1922," *University of Texas Bulletin*, February 15, 1923.

"Waco Disaster, The," *Waco Heritage & History*, Spring 1981.

Waring, Katharine T., "Ben Ficklin's Flood (Excerpts)," *Fort Concho Report*, Fall 1982.

Wheeler, David L., "The Texas Panhandle Drift Fences," *Panhandle-Plains Historical Review*, Volume 55, 1982.

_____. "The Blizzard of 1886 and Its Effect on the Range Cattle Industry in the Southern Plains," *Southwestern Historical Quarterly*, Volume 94, January 1991.

Williams, J.W., "A Statistical Study of the Drouth of 1886," *West Texas Historical Association Year Book*, 1945.

## Books

Adams, Ramon F. *Western Words: A Dictionary of the American West.* University of Oklahoma Press, Norman, 1968.

Baker, T. Lindsay. *Ghost Towns of Texas.* University of Oklahoma Press, Norman, 1986.

Baker, J.W. *A History of Robertson County, Texas.* Texian Press, Waco, 1971.

Barnard, Edward S., editor. *Story of the Great American West.* The Reader's Digest Association, Pleasantville, 1977.

Bomar, George W. *Texas Weather,* second edition. University of Texas Press, Austin, 1995.

Cartwright, Gary. *Galveston: A History of the Island.* Texas Christian University Press, Fort Worth, 1991.

*Centennial History of Corpus Christi.* Corpus Christi Caller-Times, Corpus Christi, 1952.

*Climates of Texas Counties, The.* National Fibers Information Center, The University of Texas at Austin, 1987.

*Corpus Christi Chronicles, The: A Quincentary Salute to the Americas.* Corpus Christi Quincentary Commission, Corpus Christi Chamber of Commerce, Austin, 1992.

Coulter, John, editor. *The Complete Story of the Galveston Horror: Written by the Survivors.* United Publishers of America, New York, 1900.

Cummings, Joe. *Texas Handbook,* fourth edition. Moon Publications, Chico, California, 1998.

Dary, David. *Cowboy Culture.* Avon, New York, 1981.

Dewees, William R. *Letters from an Early Settler of Texas.* Texian Press, Waco, 1968.

Dobie, J. Frank. *Cow People.* Little, Brown & Co., Boston, 1964.

_____. *The Longhorns*. University of Texas Press, Austin, 1990.

_____. *Tales of Old-Time Texas*. Little, Brown and Co., Boston, 1955.

Ellis, Michael J. *1984 Hurricane Almanac*. privately published. Corpus Christi, 1984.

Fehrenbach, T.R. *Lone Star: A History of Texas and the Texans*. American Legacy Press, New York, 1983.

Griffiths, John F., and Ainsworth, Greg. *One Hundred Years of Texas Weather 1880-1979*. Office of the State Climatologist, Department of Meteorology, College of Geosciences, Texas A&M University, College Station, 1981.

Haley, J. Evetts. *Charles Goodnight: Cowman and Plainsman*. University of Oklahoma Press, Norman, 1936.

_____. *The XIT Ranch of Texas and the Early Days of the Llano Estacado*. University of Oklahoma Press, Norman, 1967.

Halstead, Murat. *Galveston: The Horrors of a Stricken City*. American Publishers' Association, (no city), 1900.

*History of Gonzales County, Texas, The*. Gonzales County Historical Commission, Curtis Media Corp., Gonzales, 1986.

Henry, Walter K., Driscoll, Dennis M., and McCormack, J. Patrick. *Hurricanes on the Texas Coast: The Destruction*. Texas A&M University Press, College Station, 1975.

Hoyt, William G., and Langbein, Walter B. *Floods*. Princeton University Press, Princeton, 1955.

Hutto, Nelson. *The Dallas Story from Buckskins to Top Hat*. William Noll Sewell, Dallas, 1953.

*Indianola Scrap Book: Fiftieth Anniversary of the Storm of August 20, 1886*. The Victoria Advocate, Victoria, 1936.

Kelley, Dayton, editor. *The Handbook of Waco and McLennan County, Texas*. Texian Press, Waco, 1972.

Kelton, Elmer. *Elmer Kelton Country: The Short Nonfiction of a Texas Novelist*. Texas Christian University Press, Fort Worth, 1993.

_____. *The Time It Never Rained*. Texas Christian University Press, Fort Worth, 1973, 1984.

Lamar, Howard R., editor. *The New Encyclopedia of the American West*. Yale University Press, New Haven, 1998.

Lea, Tom. *The King Ranch*. Little Brown, Boston, 1957.

Leckie, Shirley Anne, editor. *The Colonel's Lady on the Western Frontier: The Correspondence of Alice Kirk Grierson*. University of Nebraska Press, Lincoln, 1989.

Leckie, William H. *The Buffalo Soldiers: A Narrative of the Negro Cavalry in the West*. University of Oklahoma Press, Norman, 1967.

Ludlum, David M. *Early American Hurricanes 1492-1870*. American Meteorological Society, Boston, 1963.

Lynch, Dudley. *Tornado Texas Demon in the Wind*. Texian Press, 1970.

McCallum, Henry D. and Frances T. *The Wire That Fenced the West*. University of Oklahoma Press, Norman, 1965.

McCarver, Norman L., and McCarver, Norman L. Jr. *Hearne on the Brazos*. Century Press of Texas, San Antonio, 1958.

Malsch, Brownson. *Indianola: The Mother of Western Texas*. State House Press, Austin, 1988.

Mason Jr., Herbert Molloy. *Death from the Sea: Our Greatest Natural Disaster: The Galveston Hurricane of 1900*. The Dial Press, New York, 1972.

Monday, Jane Clements, and Colley, Betty Bailey. *Voices from the Wild Horse Desert: The Vaquero Families of the King and Kenedy Ranches*. University of Texas Press, Austin, 1997.

Moore, Harry Estill. *Tornadoes over Texas: A Study of Waco and San Angelo in Disaster*. University of Texas Press, Austin, 1958.

Merrill, William E. *Captain Benjamin Merrill and the Merrill Family of North Carolina*. self-published, Penrose, N.C., 1935.

*New York Public Library American History Desk Reference, The*. Macmillan, New York, 1997.

Nixon, Jay. *Stewards of a Vision (A History of King Ranch)*. King Ranch Inc., 1986.

Nordyke, Lewis. *Great Roundup: The Story of Texas and Southwestern Cowmen*. William Morrow & Co., New York, 1955.

Notson, William M. *Fort Concho Medical History 1869 to 1872*. Fort Concho Preservation and Museum, San Angelo, 1974.

O'Neal, Bill. *Historic Ranches of the Old West*. Eakin Press, Austin, 1997.

Ousley, Clarence, editor. *Galveston in Nineteen Hundred*. William C. Chase, Atlanta, 1900.

Poage, W.R. (Bob). *McLennan County—Before 1980*. Texian Press, Waco, 1981.

Pruett, Jakie L., and Cole, Everett B. *The History & Heritage of Goliad County, Goliad County Historical Commission*. Eakin Publications, Austin, 1983.

Sanders, Ti. *Weather: A User's Guide to the Atmosphere*. Icarus Press, South Bend, 1985.

Santerre, George. *The White Cliffs of Dallas*. Book Croft, Dallas, 1955.

*Saragosa, Texas, Tornado, May 22, 1987: An Evaluation of the Warning System*. National Research Council, National Academy Press, Washington, D.C., 1991.

Scogin, Russell Ashton. *The Sanderson Flood of 1965: Crisis in a Rural Texas Community*. edited by Earl H. Elam, Sul Ross State University, Alpine, 1995.

Seale, William. *Texas in Our Time: A History of Texas in the Twentieth Century*. Hendrick Long Publishing Co., Dallas, 1972.

Sowell, A.J. *History of Fort Bend County*, W.H. Coyle & Co., Houston, 1904.

Spence, Clark C. *The Rainmakers: American "Pluviculture" to World War II*. University of Nebraska Press, Lincoln, 1980.

Stephens, A. Ray, and Holmes, William M. *Historical Atlas of Texas*. University of Oklahoma Press, Norman, 1989.

Taft, Harold, and Godbey, Ron. *Texas Weather*. England and May, Oklahoma City, 1975.

*Texas State Travel Guide*. Texas Department of Transportation, Travel and Information Division, Austin.

Vestal, Stanley. *Short Grass Country*. Duell, Sloan & Pearce, New York, 1941.

von Richthofen, Walter Baron. *Cattle-Raising on the Plains of North America*. University of Oklahoma Press, Norman, 1964.

*Waco Tornado 1953: Force that Changed the Face of Waco*, Volumes 1-2. Waco-McLennan County Library, Waco, 1981.

Wallis, George A. *Cattle Kings of the Staked Plains*. Sage Books, Denver, 1964.

Webb, Walter Prescott. *The Great Plains*. Grosset & Dunlap, New York, 1931.

Weems, John Edward. *A Weekend in September*. Texas A&M University Press, College Station, 1957, 1980.

_____. *The Tornado*. Texas A&M University Press, College Station, 1977, 1991.

White, Beth. *Goliad Remembered 1836-1940: The Raucous, Cultured Century!* Nortex Press, Austin, 1987.

*Wichita Falls, Vernon & Lawton Tornadoes, The*. C.F. Boone Publishers, Wichita Falls, 1979.

Wilkins, Frederick. *The Law Comes to Texas: The Texas Rangers 1870-1901*. State House Press, Austin, 1999.

Ziegler, Jesse A. *Wave of the Gulf*. The Naylor Company, San Antonio, 1938.

# Index